UNCORRECTED ADVANCE PROOF

Title: **A Certain Ambiguity:** ***A Mathematical Novel***

Author: Gaurav Suri & Hartosh Singh Bal

US Publication Date: August 2007

ISBN: 978-0-691-12709-5 Cloth $27.95T

Pages: 304 pages, 56 line illus, 4 tables, 6 x 9

For additional information or questions, please contact:

Andrew DeSio, Publicity Manager
Tel 609-258-5165
Fax 609-258-1335
andrew_desio@pupress.princeton.edu

PRINCETON UNIVERSITY PRESS
41 William Street, Princeton, NJ 08540
(609) 258-3897 Phone, (609) 258-1335 FAX

In Europe contact:
Caroline Priday, Publicity and Marketing Manager
cpriday@pupress.co.uk
Princeton University Press
3 Market Place
Woodstock, England
OX20 1SY
Tel +44 (0) 1993 814503
Fax+44 (0) 1993 814504

A CERTAIN AMBIGUITY

A Mathematical Novel

Gaurav Suri & Hartosh Singh Bal

While taking a class on infinity at Stanford in the late 1980s, Ravi Kapoor discovers that he is confronting the same mathematical and philosophical dilemmas that his mathematician grandfather had faced many decades earlier—and that had landed him in jail. Charged under an obscure blasphemy law in a small New Jersey town in 1919, Vijay Sahni is challenged by a skeptical judge to defend his belief that the certainty of mathematics can be extended to all human knowledge—including religion. Together, the two men discover the power—and the fallibility—of what has long been considered the pinnacle of human certainty, Euclidean geometry.

As grandfather and grandson struggle with the question of whether there can ever be absolute certainty in mathematics or life, they are forced to reconsider their fundamental beliefs and choices. Their stories hinge on their explorations of parallel developments in the study of geometry and infinity—and the mathematics throughout is as rigorous and fascinating as the narrative and characters are compelling and complex.

Moving and enlightening, ***A CERTAIN AMBIGUITY*** is a story about what it means to face the extent—and the limits—of human knowledge.

Gaurav Suri, a partner at a global management consulting firm in San Francisco, holds a master's degree in mathematics from Stanford. **Hartosh Singh Bal**, a leading independent journalist in New Delhi, holds a master's degree in mathematics from New York University. Suri and Bal have been friends since childhood. This is their first book.

AUGUST
978-0-691-12709-5 Cloth $27.95T - £ 16.95
304 pages. 56 line illus. 4 tables. 6 x 9.
POPULAR SCIENCE – MATHEMATICS – FICTION

A Certain Ambiguity
A Certain Ambiguity
A Certain Ambiguity
A Certain Ambiguity
A Certain Ambiguity
A Certain Ambiguity
A Certain Ambiguity
A Certain Ambiguity
A Certain Ambiguity
A Certain Ambiguity
A Certain Ambiguity
A Certain Ambiguity
A Certain Ambiguity
A Certain Ambiguity
A Certain Ambiguity
A Certain Ambiguity
A Certain Ambiguity
A Certain Ambiguity
A Certain Ambiguity
A Certain Ambiguity

A Certain Ambiguity

A Mathematical Novel

A Mathematical Novel
A Mathematical Novel
A Mathematical Novel
A Mathematical Novel
A Mathematical Novel
A Mathematical Novel
A Mathematical Novel
A Mathematical Novel
A Mathematical Novel
A Mathematical Novel
A Mathematical Novel
A Mathematical Novel

GAURAV SURI
HARTOSH SINGH BAL

PRINCETON UNIVERSITY PRESS • PRINCETON AND OXFORD

Published by Princeton University Press, 41 William Street, Princeton, New Jersey 08540

In the United Kingdom: Princeton University Press, 3 Market Place, Woodstock, Oxfordshire OX20 1SY

ISBN 978-0-691-12709-5

British Library Cataloging-in-Publication Data is available

This book has been composed in Electra LH

Printed on acid-free paper. ∞

press.princeton.edu

Printed in the United States of America

1 3 5 7 9 10 8 6 4 2

10 9 8 7 6 5 4 3 2 1

To Gaurav's parents, Ramesh and Surya Kanta Suri

To Hartosh's parents, Neena and Amarjit Singh Bal

Author's Note Author's Note Author's Note

The narrative you are about to read—*A Certain Ambiguity*—is a work of mathematical fiction. Its mathematics seeks to be complete in itself and remains as substantially true in the world of fiction as it is in our world. The narrative's story line, on the other hand, is entirely fictitious and all the characters are a product of our imagination. But there is a third, fuzzy grey area between the factual mathematics and the fictional narrative: in an effort to flesh out the life and blood struggle of the mathematicians who actually did the mathematics presented in this book, we have used the device of a journal entry in which mathematicians through the ages present their experiences and mathematics in what purports to be their own voice.

While the mathematics in the journal entries has always been correctly attributed, the experiences described therein come with a large dose of literary license and are not intended to be historically accurate accounts of the events they portray. Further, much of the mathematical notation and terminology used in the journal entries has been developed in the modern era and would have been unknown to several of the mathematician "authors" themselves. However, wherever it has been possible to draw upon historical fact, whether through the written work of the mathematician concerned or through popular anecdotage, we have done so. It is in this sense that the journal entries straddle the gulf between fact and fiction.

In order to ensure that the reader does not carry away any misapprehension about what is fiction versus actuality, we have included a notes section at the end of the book that spells out the key historical facts pertaining to each journal entry. In a few cases we have also used the notes section to further explain a mathematical or philosophical point not fully developed in the text.

The Hindi word *Bauji* frequently used in the narrative is a form of address for one's grandfather.

Finally, the New Jersey Blasphemy Law mentioned in Chapter 2 takes language directly from the state statute. The last conviction of this law occurred in 1886 in Morristown, NJ. The town of Morisette, NJ appearing in the narrative is fictional.

A Certain Ambiguity
A Certain Ambiguity
A Certain Ambiguity
A Certain Ambiguity
A Certain Ambiguity

A Certain Ambiguity

A Certain Ambiguity
A Certain Ambiguity
A Certain Ambiguity
A Certain Ambiguity
A Certain Ambiguity
A Certain Ambiguity
A Certain Ambiguity
A Certain Ambiguity
A Certain Ambiguity
A Certain Ambiguity
A Certain Ambiguity
A Certain Ambiguity
A Certain Ambiguity
A Certain Ambiguity

CHAPTER

1 2 3 4 5 6 7 8

YESTERDAY I FOUND THE CALCULATOR my grandfather gave me on my 12th birthday. It had fallen behind the bookcase and I saw it when I was rearranging the study. I had not thought about it for years, yet when I held it, it seemed as familiar as ever. The "I" in Texas Instruments was missing, as it had been for all but two days of its life; the buttons still made a confirming clicking sound when pressed; and when I put in some new batteries, the numbers on the LCD shone through with a blazing greenness, more extravagant than the dull grey of the modern calculator. My grandfather had intended this calculator to mark a change in my life—a new direction. As it turned out, it did mark a change, though not the one he had in mind.

I punched in the number 342 without thinking about it. It was the same number I had entered 25 years ago when the calculator was brand new.

"Want to see some number magic?" my grandfather had asked as he watched me push the buttons more or less randomly. I was sitting in his room completely taken by his birthday present, if not quite sure what to do with it. He put his notebook down, temporarily giving up on the math problem that had resisted solution since morning.

"Yes, Bauji!" I had rushed over to him.

"Enter any three-digit number in your calculator and do not let me see it." That is when I had first entered 342, the same three digits I entered now. "OK. Now enter the same number again, so you have a six-digit number," he had said. I punched in 342 again, so now I had 342342 entered in my calculator. "Now, I do not know the number you have in there, Ravi, but I do know that it is evenly divisible by 13."

By "evenly divisible" he meant that there would be no remainder. For example, 9 is evenly divisible by 3 but not by 4.

Bauji's claim seemed fantastic to me. How could he know that my number, randomly chosen and completely unknown to him, would be evenly divisible by 13? But it was! I divided 342342 by 13 and I got 26334 exactly, with no remainder.

"You're right," I said, amazed.

He wasn't finished, though. "Now, Ravi, I also know that whatever number you got after you divided by 13 is further divisible by 11." He was right once again. 26334 divided by 11 was 2394. Why was this working? "Take the number you got and divide by 7. Not only will it divide evenly, but you will be surprised at what you get." He had begun his pacing and I knew that he was as excited as I was.

I divided 2394 by 7 and I got 342! "Oh! Oh! It's the number I started with! Bauji, how did this happen?"

My grandfather just sat there, grinning at the completeness of my astonishment. "You will just have to figure that one out Ravi," he said, walking out to check the state of his tomato plants, the newest additions to his vegetable garden in the backyard. He seemed to be the only person who could grow tomatoes in New Delhi's dry summer heat.

The first thing I did was to check the divisions by hand. My hypothesis was that Bauji had rigged the new calculator somehow. But no, the numbers worked out exactly the same way when I did the long divisions by hand. Next, I decided to try this with some other three-digit numbers. The same thing worked every time. Whatever the repeated number, I could divide it evenly by 13, 11, and 7, and each time I got back to the number I started with. A few minutes of checking and rechecking convinced me that this property was true of any three-digit number. I tried doing the same thing with four-digit numbers, and it did not work any more. Neither did it work for two-digit numbers. What was going on?

I tried reversing the order. Instead of dividing first by 13, then 11, and then 7, I divided the six-digit number first by 7, then 11, and then 13. It made no difference at all. After dividing by each of those three numbers I would get back my original three-digit number. Why was this happening?

I wasn't getting anywhere and it was getting to be dinnertime. Ma had already called me twice. I knew that risking a third "I'll be right there" would be unwise, and so I put my notebook away and headed to the kitchen. I stopped thinking about the problem. And then mysteriously, out of nowhere, just when I was wondering if Ma would let me have ice cream, a new idea occurred to me. Even now, with more experience in such things, I cannot quite explain the inception of the moment of insight—the Aha moment—when out of nowhere a new idea comes, and chaos is replaced by understanding.

My first real Aha moment was at the dinner table, two days after my 12th birthday. The idea that loosened the knot was the realization that division was the reverse of multiplication, a fact I had long known, but never applied in quite the fashion this problem demanded. Instead of dividing by 13, 11, and 7 one at a time, why not multiply them together and then divide by the product all at once? Would this approach even lead to the same answer? I thought it would, and a confirming example showed this to be the case. In my head I divided the number 24 first by 2 and then by 3. I got 4 as the answer. Next I divided 24 by 6 (which is 2×3) and got 4 as well. So, it should work to divide the six-digit number by the product $13 \times 11 \times 7$. I did some more examples on a paper towel just to be sure. It appeared that I might be onto something.

Ma noticed that I was completely distracted. "Ravi, what's going on? Why aren't you eating?" But I hardly heard her. I had to find out what $13 \times 11 \times 7$ was; perhaps that would lead me to understand why Bauji's magic worked.

"I'll be right back, Ma," I said, getting up quickly before she could react.

"No, sir, you won't. Sit here and finish your dinner." She looked like she meant it.

But my grandfather must have known what I was going through.

"Its okay Anita. Let him go." He must have been convincing enough, for I saw unwilling permission in my mother's eyes.

I ran up the stairs two at a time and fired up the calculator. $13 \times 11 \times 7$ was . . . 1001.

I knew this was terribly significant, though as yet I was not quite sure why. I tried dividing 342342 by 1001. As I expected, I got 342. But wait a minute. That must mean that the reverse is true as well. So if I multiply 342 by 1001 I should get 342342. Of course! $342 \times (1001) = 342 \times (1000 + 1) = 342{,}000 + 342 = 342342$. So, taking a three-digit number and repeating it was just like multiplying it by 1001. And if you multiplied it by 1001, you could divide the six-digit number by 1001 to get the original three-digit number. What had confused me was dividing by 13, 11, and 7—but by dividing by those numbers I was in effect dividing by 1001. How simple! How could I have not seen it?

"Bauji! Bauji! I've got it!"

And he too bounded up the stairs two at a time, just as I had a few minutes before. "Tell me," he gasped. He was out of breath, but not overly so—not bad for 85.

"When you asked me to repeat the three-digit number, you were actually having me multiply it by 1001. And then you made me divide it by 1001, except you did it in three stages. So of course I ended up with the same number I started with!"

He looked at me and smiled. "Good work," he said ruffling my hair, his most characteristic gesture of affection. "I'll give you another one to think about tomorrow." When I told him I wanted another one right then, he laughed. "Looks like you'll be the next mathematician in the family. We might have to send you to The Institute of Advanced Studies! Maybe we could collaborate on some research."

Sitting in his lap, surrounded by his books and papers, I could not imagine a better fate.

• • •

The next evening Bauji died. I remember going to his room, calculator in hand, ready for my next puzzle, but as I neared the door I heard my mother's voice—really a whisper—coming from inside the room. The door was ajar, about half open. I could hear an urgent pleading in her tone even though her sounds did not seem to have the rhythm of words. From the hallway I saw that she sat on the floor, cross-legged, near my grandfather's desk. She had Bauji's—her father's—head in her lap and she was bent over him massaging his forehead, beseeching him. Even though I had never seen a dead person before, and even though I was not standing near him, I could tell that Bauji was gone. His posture had an unalterable finality that sleep lacks. For many minutes I didn't move. I stood in the doorway of his room, which suddenly seemed extraordinary in its sameness: his desk was piled high with its usual mountain of mathematics books; three of them were open. On the far wall his books were spilling out of the two large bookcases; many were on the floor. On the wall near me were his music records and cassettes, mostly instrumental jazz. The tape-recorder was set on auto-replay and was softly playing Louis Armstrong's trumpet rendition of "Summertime."

At the sound of my footsteps Ma lifted her head and her eyes met mine. For a second I saw her face contort, her eyes closed tightly on themselves and tears flowed from the corners. She quickly pulled me towards her so I wouldn't see her cry. From over her shoulder I could see Bauji's last expression. He wore the face of happy surprise, as if he had at last glimpsed the solution to a difficult mathematical problem, and the answer was not at all the one he had expected.

As my mother hugged me more tightly the calculator slipped from my hands and fell to the floor near Bauji's hands. The "I" of Texas Instruments fell out and could not be reattached despite my, and then my father's, best attempts.

Many years later my mother told me that Bauji had decided that I had a mathematician's mind and he had wanted to push me to excel in the subject. The calculator was to have been a catalyst in my development as a mathematician. "I'm going to use it to get Ravi passionate about mathematics," Bauji had said.

• • •

Although Bauji was secular, he participated in religious functions with some regularity. "Religion is about community," he would announce after each such function, "and everyone needs community." I once heard an uncle refer to him as an "atheist with goodwill towards God." I didn't know quite what he meant, but it somehow seemed to fit Bauji.

So when he died, no one was quite sure what type of ceremony was required. He would be cremated, that much was clear. All Hindus are cremated, and while Bauji never referred to himself as a Hindu, he never repudiated the affiliation either. But the family divided on the extent of the rites that should accompany the sacrament. Bauji's sister insisted upon a full recitation from the scripture, sprinkling of holy Ganga-jal, spreading of gold dust, application of sandalwood paste, the presence of six Brahmans, and the lighting of the pyre by the eldest male descendant.

My mother disagreed. "Bauji liked simplicity, he would not have wanted all this," she said. After much wrangling (that got her crying) Ma prevailed on every point except the lighting of the pyre by the eldest male descendant.

"There is no other option. It has to be a male descendent or his soul will not be properly liberated. It says so in the Vedas," insisted my great aunt, and on this one issue she refused to give in. And since Bauji had no sons and I was the only grandchild, the eldest male descendant they were talking about was me.

When I reached out with the kindling to light the funeral pyre, my hand started to shake. The shaking was strong, and it grew stronger when

I unexpectedly started to shiver: I was embarrassed: I had wanted to present an image of dignified grief to my relatives—so becoming in a 12-year-old faced with the passing of his grandfather, whom he so adored—but it wasn't working. I had the odd sensation of watching myself from outside my own body, as if I were floating around the entire ceremony, watching it from above. I could see the boy in the middle who now (as a final embarrassment) appeared to be crying, who was losing the battle to still his hands, which were stubbornly refusing to obey his commands. Then I saw my father hold the boy from behind and steady his wrists. At his touch, I came back into myself. The flame caught. The pyre hissed.

• • •

A few days after all the relatives left I got in the habit of going to Bauji's room every afternoon after school. I would lie down on the floor in the center of the room and imagine he was still sitting on his desk by the window. When he was wrestling with a problem he would sit there with his eyes shut and his body perfectly still. He would stay like that for a long time and then, every once in a while, he would sit up very straight and furiously start writing in his notebook. If he liked what he wrote, he would jump up, as if released by a spring, and pace with great energy and intensity, muttering to himself, or sometimes to me, "Could this be it? Could this be it?" Sometimes he would end these walkabouts with a loud "Ha!" and take me in his arms and throw me high up, nearly to the ceiling, and then catch me under my armpits as I came down. "I see it now, Ravi! I see it!" Now and then he would challenge me to have a go at a mathematical question, and, if I succeeded, he and I would do a joyful postmortem on the insight that cracked the case.

Bauji saw grace in mathematics, and sometimes I could see glimpses of what he saw. Now, without him, there was only the monotonous drone of doing what needed to be done.

I did try to read his mathematics books, but their pages seemed cold and lifeless. The symbols spread themselves on page after page without reason or beauty. I looked at his handwritten notebooks and they too had an alien feel, except for one page whose margin contained the notation "Show this to Ravi" next to some ominous-looking calculations. For two days I tried and failed to decipher what he might have wanted to show me. I could only tell that it seemed to have something to do with prime numbers and infinity.

After I was beaten by his math, I took to going to his room and listening to his jazz records. At first they, too, seemed without order, like the mathematical symbols in his books. There was none of the predictable, repeating structure of the music I was accustomed to; instead there were notes that seemed floated on the spur of the moment, without a planned arrangement to guide them.

Then suddenly one day, I got it. I was listening to Charlie Parker (*Crazeology*) and I had a musical Aha moment. I realized that in most of these records there was, in fact, an underlying structure which allowed for inspired improvisation: the ensemble would first play the tune from beginning to end, with the melody played by the horns, and the harmony played by the rhythm section—the piano, bass, and drums. Then, as the tune went on, the rhythm section would continue to play the harmony while each horn improvised a solo. The soloist would select notes available within the harmonic structure while incorporating the soul of the original melody, but with his notes he would create something new each time he played. And it was all done with a casual, understated coolness. Within two weeks I was hooked.

So I spent the afternoons and evenings listening to Parker, Armstrong, Ellington, and then Goodman and Getz. For two months my parents let me be. They must have heard the music coming from Bauji's room—the music that everybody except Bauji thought was strange—yet they did not ask me about it.

It was not until a week before my final examinations that my mother came to institute some course correction. I was sprawled on the floor, my eyes shut and my feet keeping pace with the changing moods of "West End Blues." She tapped me on the shoulder and asked if she could turn off the record player. "We need to talk," she said. I could tell she was picking her words carefully. She told me she knew how much I missed Bauji and she knew I was listening to his music to "stay connected with him." I wasn't sure of this—I thought I was listening to his music because I liked it—but I did not think it best to volunteer this information. "Ravi, it's time to move on. Bauji wanted you to follow a path in life and you can't get on that path if all you do is listen to this . . . music."

Then she gave me the big news. "Bauji has left you a lot of money. Most of his life savings, actually. His will says that you must use this money to go to college in America. And you can't do that unless you keep doing well in school. You have one week to rescue your grades. If you want to go, you have to bear down—not just this week, but all the way through high school."

America. Land of freedom. Land of Louis Armstrong, The Institute of Advanced Studies, and wide open roads that could take you anywhere. Of course I wanted to go. But more importantly, Bauji wanted me to go.

So I studied hard. Over the next six years I became a repository of facts about accounting rules, thermodynamics laws, Sanskrit verb types, inorganic compounds, and different rock structures. My grades were excellent, my capacity to store information phenomenal. But it was a joyless endeavor. I excelled in all subjects but saw beauty in none. Even mathematics lost its luster; it became like a game with well-defined rules to be followed for the sole purpose of acing examinations.

But get the grades I did, and one Friday evening just before my eighteenth birthday, I got a letter from Stanford University inviting me to come and study there.

• • •

The first person I met in California was Peter Cage. Peter had volunteered to pick up international students from the airport and drive them to campus. He greeted me with a wide smile and, unexpectedly, a hug. "Welcome to America," he said, appearing to mean it. When we were in his car (a newish Toyota), I asked him what motivated him to volunteer to help foreign students. "Looks good on the resume," was his answer. "A multinational company might look upon this experience very favorably." He said this without guilt or apology, and I instantly liked him for it. I knew others who signed up for causes to get credit of one kind or another, but invariably they would ascribe their volunteerism to a higher calling. Not Peter. He and I became friends, then roommates.

Peter was stubbornly wholesome. He ran three miles every morning, ate right steadfastly, kept his room orderly, (all his books were always stacked; mine never were), did his homework on time, and was never noticeably down. Even more incredibly, he seemed to be singularly free of any doubt. He was majoring in business because it was the best way to get into investment banking; he would practice banking because it was the best way to get rich; he would get rich because it would lead to freedom. And this wasn't all talk either. Now, 19 years after our first freshman semester, Peter is one of the leading investment bankers in the technology sector in Silicon Valley. He decided what he was going to do, and then he did it.

I, on the other hand, was filled with doubt. I had difficulty getting interested in any subject. Getting good grades was not really an end in itself anymore because, in my mind, my contract with Bauji was fulfilled the day I was admitted to Stanford. And without the grade imperative, I drifted. I had brief flashes of interest in astronomy, Roman history, and game theory, but nothing really took. I had great difficulty choosing a major. It was not until the second semester of my junior year that I finally did so—and that, only at my father's urging. He thought that economics would make me attractive to a wide variety of corporate recruiters. Having no vision of my own, I went along with his.

Peter, too, was enthusiastic about my choice of major. "You can't go wrong with economics. You can do anything with it, even investment banking." But the thought of dedicating my life to any one thing seemed too heavy to me. Whatever I picked, how could I be sure that I had picked well? How could I be sure that what I was going to dedicate my life to was worthy? This lack of certainty became a theme for me. The big choices of career path and specialization had laid the seed, but gradually even lesser things, such as which class to take or which book to read next, caused internal debate. I wanted some way to know that my choice was right.

The one event free of any such internal strife was "Thursday Night Jazz," a student jam session where anyone could come and perform. It typically started at 11:00 p.m. and went on 'til 2:00 a.m., or until everyone who wanted to play had gotten a turn on stage. The best musicians were allowed to play early—before midnight—while there was still an audience to be had. After midnight most everybody left, and the only people remaining in the audience were other musicians who were yet to play. Except me. I would come early and, more often than not, stay till closing, even though I came to listen, not to play. I did play once—it was during my sophomore year—when I allowed beer and the enthusiastic goading of Peter Cage to overcome my self-consciousness and banged out an insipid "The Way You Look Tonight" on the piano. But truth be told (as Peter frequently reminded me), I was no worse than 90% of the people who got up there; I just had a harsher internal critic.

But my critic was more sympathetic when it came to other people. I easily tolerated their mistakes. I told myself that I didn't come to listen to jazz perfection—a Miles Davis CD was all anyone needed for that. I came to listen to live performances that, though flawed, were more immediate and powerful than any recording (both for the player and the listener). With few exceptions, every person who came up on stage had heard the beauty of a perfectly executed improvisation. The fact that the tones playing in their heads were somewhat different from the ones coming out of their instruments was unfortunate, but in my view did not negate the nobility of their attempt. Despite all the botched harmonics and the poor timing, there were still moments of beauty. You just had to wait for them.

By the beginning of my senior year I had become a recognized regular and knew many of the musicians who played frequently. I would sit at the same spot, and if I was alone I'd carry along whatever I was reading. Sometimes Peter came with me, but he would always leave by midnight. "I need to get up early," he'd say without pride or regret. But one Thursday just before the first semester of our senior year, because classes were yet to start, he made an exception and stayed late. And that was the night I met Nico Aliprantis. It was Peter who pointed him out. "See that guy over there? He's the best math teacher I've ever had," he said.

At that time Nico was 62, which made him approximately three times as old as most people in the room. But he fit right in. His walk was tall and easy, his manner comfortable, and his mouth always on the verge of an amused smile. He chose a table near the stage, put his motorcycle helmet aside, and proceeded to roll his own cigarette.

"What class did you take from him?" I asked Peter. To date, I had been unaware that he cared about the quality of math teachers.

"Statistics," he said. "He was the only teacher that ever made math seem like a natural thing, not just a bunch of rules."

Nico listened to the music attentively. From time to time someone (probably a former student) would stop by his table and say hello. Most professors would have rated a brief nod and that too only if there happened to be some accidental eye contact, but in his case there seemed to be a reservoir of genuine goodwill. On two occasions a student pulled a chair up to his table and stayed for a chat.

Just before closing Nico went to the stage and asked for the saxophone. He played an old Charlie Parker tune whose name I could not place, but I had heard it before; it was one of Bauji's favorites. After establishing the refrain, he began to improvise, and I knew within a minute that he was good. He played effortlessly. He knew how to get from note to note seamlessly, with a light touch and his own unique style that he somehow intertwined with Parker's. You heard Charlie Parker but you also heard Nico Aliprantis, and the two coexisted with ease. Towards the end he got tangled up and lost his way. His eyebrows squeezed together and his forehead wrinkled, and for a second he looked angry with himself. But then he decided to finish and played a nice sequence to bring the tune to a logical conclusion. He bowed and everyone clapped—some, because he was different from the rest, and others, because he was good.

After everyone had played and the Coffee House was closing down, Peter and I caught up with Nico. "Dr. Aliprantis, you were fantastic!" said Peter.

He smiled and looked at us, recognizing Peter. "You've taken one of my classes," he said, peering at him from behind his glasses. And then after a second, "Peter Cage, right?"

I was surprised he remembered—he must have had hundreds if not thousands of students. But then he said, "You were great in that statistics class. I kept saying you should study mathematics instead of business," and I understood then that Peter had distinguished himself enough to be memorable.

"Dr Aliprantis, this is my friend Ravi Kapoor," Peter said, turning towards me. As I shook his hand I told Nico that I recognized the Charlie Parker tune, but couldn't recall the title.

"'Now's the Time'," he said, looking at me more closely. "You must know jazz because that's not one of Bird's most famous recordings."

Before I could reply Peter jumped in. "Ravi knows a lot about jazz."

Nico smiled. "Do you play?"

"Not well; otherwise I'd do it as a career," I said.

Nico nodded, earnest for the first time. "I'm the same way," he said. "This math gig was a fallback choice, though fortunately I love the subject and I'm good at it, much better than I am at jazz anyway."

"You were good," I told him. "That was a great sequence you created and it worked perfectly except for that little bit at the end."

He shook his head, "I may be good compared to some guy on the street, but I'm no Charlie Parker." He said it so matter-of-factly that there was nothing further for Peter and me to say. We stood there in silence for a few seconds and then I saw Nico notice the strain and make a conscious decision to steer the conversation away from himself. "So what else do you like besides jazz?" he asked me.

Nothing really, was the truth. "I used to love mathematics," was what I came up with instead.

"Used to?" Nico asked. He asked so softly and with such benevolent curiosity that I found myself telling him the truth.

"My grandfather made mathematics inspiring and fun. I've never had anyone else who could enthuse me the way he could."

Nico smiled at that. "There's a challenge!" he laughed. "Listen," he said arriving at a decision, "Why don't both of you sign up for the class I'm teaching this fall? It's called 'Thinking about Infinity'. It's Math 208, I think. You should check it out; it should be an interesting class. We start Monday."

Walking home that night Peter and I talked about whether we should accept Nico's invitation. Peter already had an offer from Morgan Stanley that he was going to accept. He was one of the few people who had a job offer at the beginning of senior year—most people got offers later, typically in the fall semester. Without the pressure of the job hunt, he had the luxury of experimenting with "fun" classes and thought that Nico's class would fit the bill. I, on the other hand, had declared my major only a semester ago and needed to take five economics classes to graduate on schedule.

"It doesn't make sense for you, though," observed Peter making the same calculations I had just gone through. "You have to be Mr. Economics this semester." I knew he was right.

But that night, just before falling asleep, I decided to sign up for Nico's class after all. I knew this would mean having to take (a nearly impossible) *six* economics courses next semester or else taking a class in the summer, which would be a huge financial strain on my family (they were augmenting Bauji's bequest). But there was something about Nico.

• • •

There were about 15 students who showed up for "Thinking about Infinity." Peter, as was his custom, had arrived early and found a spot in the front row. The other faces seemed unfamiliar save for a slender, curly-haired saxophone player I had seen a few times at Thursday Night Jazz. His music had not been memorable, but for some reason his name had stuck with me: Adin something. He sat in the front talking to Peter, who evidently knew him from somewhere.

Nico entered the room with the same languid ease that he had shown entering the Coffee House. "Good morning everybody. My name is Nico Aliprantis and we're going to spend this semester using our finite brains to think about infinity." He smiled at his own line. I wondered if it was improvised.

He quickly went over the logistics: we would meet once a week for the next 10 weeks, each class would be three hours long with a 10-minute break in the middle, office hours would be Wednesday afternoon, and the grades would be based on two take-home tests and the quality of class participation. No prior mathematics was required—This was a mathematics course for liberal arts majors. No textbook was required either; he would hand out notes when necessary. A student asked how people were supposed to study without a textbook.

"You'll see," Nico replied without disguising his sigh. He must have gotten this question all the time. "I think you'll find that attending class and thinking about the problems I present to you from time to time will provide all the structure you need."

He took off his glasses and faced the room. "There are two themes that are going to run throughout this course. I want to talk about them up front so that as we go into the subject matter you know what to look for. First, if you allow yourself to, you will find great beauty here. I think that mathematics is beautiful at its core; it is much more like a musical piece than an accounting formula." He looked up to see how the class was receiving this idea, and that's when he happened to catch my eye. "Much more like a jazz piece," he said with a half-wink. "G. H. Hardy, a famous English mathematician, said that good mathematics is about making good patterns. A painter makes patterns with shapes and colors, a poet with words. A mathematician makes patterns with ideas." He said "ideas" loudly and the word seem to reverberate in the silence that followed.

After a minute or so in which no one said anything, I could tell that Nico was scanning the room looking for someone to talk to. He settled on Adin. "You, sir," he said pointing, "what is your name and what do you study?"

Adin's deep voice did not match his slender frame. "Adin Kaminker. I'm majoring in philosophy," he said.

"Adin, do you have a favorite poem or song?"

"Sure," said Adin. "I quite like poetry actually."

"Excellent. Okay, may I ask you to recite a few lines from a poem that you find particularly beautiful?"

Adin did not hesitate. He picked an old favorite of Bauji's. "The woods are lovely, dark and deep / But I have promises to keep / And miles to go before I sleep / And miles to go before I sleep." His recital was practiced and smooth. "That's by Robert Frost," he concluded.

The class collectively turned towards Nico, their heads moving in unison right after Adin finished his recital, like a gallery watching a tennis match. "Thank you, Adin," Nico said, bowing his head in appreciation. "You recited the lines beautifully."

He looked up, addressing the whole room now, not just Adin. "Now let's imagine Robert Frost writing those lines. Maybe he played around with which words to use—perhaps at first he used the word 'forest', instead of 'woods'. Maybe he tried many different word-sequences in many different rhythms until he got this one. And when he did, you can bet he knew that he was onto something, that he had created something beautiful. As soon as he had those lines, I'm sure he knew that they were right. They appealed to his sense of aesthetics."

Nico was pacing. He was into it. "Mathematics is done in the same way," he continued "Most mathematicians have an aesthetic sense that guides them toward the problems they try to solve and in the ways they approach them. They try many things and then, sometimes seemingly out of nowhere, an idea comes. The idea simplifies everything, puts everything in harmony. And when they have the idea they often know that they are right, even though they have not worked out all the details. With practice they get an aesthetic sense, not unlike a poet's I imagine."

A hand shot up in the back row. It was a goatee-wearing guy in beach flip-flops, shorts, and a longish—but surprisingly disciplined—pony tail.

"Your name please?" asked Nico. It turned out that unlike most teachers, Nico had a good memory for names.

"Percy Klug, but most people call me PK."

"Go ahead, PK," he said.

"If mathematics is so beautiful, why haven't I ever heard anyone talk about it that way before?"

He was right. Mathematics was seldom seen to be beautiful. Bauji saw it that way, but he was the only person I knew who held that opinion—until now.

"I'm not sure," said Nico. "Maybe it's because mathematics is not a spectator sport. You have to do it to appreciate it, and doing it requires patience and persistence. You can love a song without being able to sing, but that doesn't work in mathematics. Nevertheless, the beauty is there for you to find." He took a sip from his coffee mug, making a slurping noise. "So the first theme is beauty. Keep a look out for it. It's not really unique to this class; I find a lot of different branches of mathematics to be beautiful. But the second theme, I think, is especially true for us. This class is also about understanding how humans think and understanding the limits of what we can think." Nico paused and looked outside towards the courtyard. When he spoke again his voice was softer and more distant. "The story of infinity is a story of how far the human mind can take us. But it is also the story of boundaries that we may not cross, no matter what. We will see amazing facts that must be true but also raise tantalizing questions that seem to be unanswerable. Not because mathematicians just happened not to have found an answer so far, but rather because they couldn't possibly. Our current set of assumptions about infinity are not strong enough to lead to an answer to some questions. *Ever.*" I didn't understand all the things Nico said but was captivated by the way he said them—like a man of faith expressing reverence in a place of worship. There was motionless silence in the ensuing pause. Then I saw Adin fish out his notebook and write something down. His pencil sounded surprisingly loud.

"You'll see what I mean as the class progresses," said Nico, coming back to us. "But let's get started today by recalling our first memory of infinity. What made you think about infinity for the first time?"

PK the surfer guy raised his hand immediately. "Space," he said. "I grew up in the desert, and at night you could see the Milky Way and it was impossible not to think of infinity when you saw all those stars."

Nico nodded and wrote "Space" on the chalkboard. "Who else?" he asked.

A Chinese woman volunteered "time" because it kept on passing. "TIME" went on the list as well.

Peter said "God," feeling the need, in our secular times, to shrug his shoulders somewhat apologetically. In his later years Peter would become more certain about his faith.

"It is hard to imagine a finite God!" nodded Nico, adding the almighty to his list. "Counting," I volunteered. When I was five, I used to play a game with Bauji of naming larger and larger numbers. Invariably I'd find myself adding 1 to whatever strange number Bauji came up with.

"Yes, of course," said Nico. "Thank you Ravi." I was surprised he remembered my name from the other night.

After a pause Adin raised his head. "For me it was space—not in the unlimited sense, but in the sense of it being unendingly divisible. I first had that thought when my parents presented me with a microscope."

"That's right, Adin. Infinity has a dual aspect, the infinitely large and the infinitely small."

Nico's list read:

Space, without bound

Time

God

Number (counting)

Space, unendingly divisible

He looked at it for a few seconds. "It's a good list," he said, his back towards us. "In each of these examples we are observing a finite object or process and extrapolating it without limit. Where there are a billion stars there could be an infinite number; time keeps passing, so it may pass without end, forever; God almost by definition must be infinite—his powers are an unending extrapolation of our finite ones; numbers do go on and on and on; and where we divide once, we could, at least in theory, divide again. By our ability to generalize and extrapolate we force infinity to exist, at least in our minds. Its existence is an affirmation of the human power of reasoning by recurrence."

"But does infinity really exist?" asked Adin. "I mean, do we know if anything on this list is actually infinite?"

Nico shrugged. "Some people say that space is unbounded but finite, that time has a beginning and an end, that God does not exist, and that numbers are only a product of the human mind. So according to this view there is nothing truly infinite in the physical universe."

"How can space be unbounded but finite?" PK wanted to know.

Nico laughed. "Good question. Perhaps space is like our planet. The earth is an unbounded surface. No matter how far you go, you'll never come to the edge. But the earth is also finite. So unbounded but finite things are certainly possible."

PK was not buying it. "That's because the earth has a flat, two-dimensional surface that curves upon itself in the third dimension to make a ball. But space is already three-dimensional; it has nothing to curve into!" PK was smarter than I had initially thought.

"Some people believe that there is a fourth dimension that we are unable to perceive. Perhaps the universe curves into the fourth dimension," said Nico.

Adin raised his hand. "There might be an infinity of dimensions then. Why stop at four?"

"It's possible, and then we could have another type of infinity, but we're only speculating here."

Peter, never one for science fiction–type theories, took us back to God. "Doesn't God have to be actually infinite in some sense?"

"If there is such a thing as God." It was Adin who replied, not Nico. Peter shrugged his shoulders without looking back. Peter seldom argued unless he thought he had a shot at changing the other person's opinion. Philosophical debates did not excite him.

Nico summarized where we were. "What we're seeing here is that there is no proof that infinity exists in nature. It may or it may not. But because numbers exist as an idea in the human mind, infinity must also exist in the human mind. If we acknowledge the existence of the number 1 and acknowledge that we can always add 1 to any number, we automatically acknowledge the concept of infinity. Any doubters?" He asked with curiosity, not with the intent to challenge. I thought that Adin was going to say something, but on due consideration he apparently found Nico's statement to be airtight. "Very well. Since infinity exists, if not in nature, then at least as a valid idea in our minds, the first thing we ought to do is find a symbol for it. John Wallis, an English mathematician, did this in 1655. Most of you have probably seen it before. It's called the unending curve."

Nico drew the symbol "∞" on the chalkboard. "Now that we've got a symbol for it, we need to try to get a better handle on what it is." He looked up at us. "That, ladies and gentlemen, is much harder than you might suspect. In fact, it is much easier to say what infinity is not. For example, we can be sure that infinity is not a number, in the sense that 943 is a number."

"Why do you say that?" asked Peter.

Nico took a piece of chalk and wrote:

$$\infty - 1 = \infty$$

"If infinity was a number it would have to be its own predecessor. If you grant me that the only types of numbers are finite numbers and infinity, observe that 1 added to any finite number cannot give infinity, so infinity minus 1 must equal infinity. But if we were to treat infinity as we treat any other number, we could subtract ∞ from both sides and deduce that $-1 = 0$, which is absurd. So infinity is not a number and may not be treated as such."

"So then what is it?" asked PK.

"That's a tough question PK," said Nico. "The Greeks tried but couldn't answer it. And despite their discovery of zero the Hindu and Arabic mathematicians couldn't come to grips with infinity either. At one point, the Hindus defined infinity to be 1/0, but then wiser heads prevailed and they realized that it cannot make sense to divide anything by 0. Most of the medieval voices either repeated Greek ideas or made infinity into a theological issue and failed to make progress. It was not until very late in the 9th century that Georg Cantor came up with a framework that made sense of infinity."

Nico had a poster of Cantor, which he now unfurled. "This man," he said pointing at the photo, "was a genius in the true sense of the word. He is the hero of our story. He single-handedly created the mathematics of infinity. Cantor defined infinity. In fact, he defined many infinites, and we'll get to his precise definitions in due time. His thinking and methods are an important focus for us in class."

What grabbed me first about Cantor's face in Nico's poster were his eyes. They looked past the camera, focusing at nothing, but strained in thought. I wondered if Cantor had been wrestling with some mathematical problem at the precise moment when the picture was taken. Only Cantor knew, and he was dead.

The bridge of Cantor's nose in the picture was straight and narrow—a Sherlock Holmes nose if there ever was one. His mouth was surrounded by a short beard that did not appear to have been trimmed carefully. It was dense in places and spotty in others. Despite the beard, I could make out the tension and the worry in his thin lips. His forehead was broad, and his scalp hairless. The photograph was grainy around the top of his head and gave the appearance of bubbling fizz on the surface of a freshly poured Coke.

"Cantor is most remembered for establishing the subject of set theory, the topic of this class. In doing this he single-handedly changed mathematics." Nico said this while looking at the photograph and slowly rubbing his chin. In the pause I felt that he would have loved to talk with Cantor in person, and frankly I would have loved to listen in on that conversation. Then, with a palpable gear shift, he turned and faced the classroom once more. He stood up straighter and his tone was firmer. It was time for mathematics.

"At an intuitive level a set is simply any collection of objects. Let me write out a few examples." He turned to the board and wrote:

A = {chair, elephant, tomato}
B = {16, watch, book, 23.75, saxophone}
C = {Godzilla, {A}}
N = {1, 2, 3, 4. . .}

"As you can see a set can have any object as a member. It is typical to collect the objects of a set within curly brackets." Nico pointed to the "{"and"}" which marked the opening and closing of his sets. "A tomato is an element (or a member) of set A, and the set A has three elements. The set **N** has an unlimited number of elements. It is not a finite set. *Infinity is simply defined as the order of a set that is not finite*."

It seemed a somewhat circular description to me, and I wasn't sure I saw the benefit. I looked up to protest but saw Nico looking at the class with an amused expression. He had anticipated our difficulties. "I can see from your faces that this definition is not the least bit satisfying. 'What is the point?' you all seem to be asking. You will see the point, I promise. More satisfying definitions of infinity require more mathematical machinery than we have at this stage. I ask you to keep this definition in the back of your mind, for it will allow us to make progress and build an amazing structure to deeply understand the nature of infinity. It is a structure that still gives me goose bumps," said Nico without any air of pretense that I could detect. "We'll get to Cantor in due time. I put the definition up front because it seems odd to begin a class about infinity without defining it. But for now, let's stay with the Greeks."

"The first Greek we'll meet is an odd bird by the name of Zeno, sometimes known as Zeno of Elea. He lived around the fifth century BC. He is said to have been a self-taught country boy. Zeno described several paradoxes built around the divisibility of space. The famous philosopher Plato dismissed these paradoxes as 'youthful efforts', yet he did nothing to resolve them. In fact, none of the best minds of the last two and a half thousand years could resolve the paradoxes raised by Zeno. Not bad for a country boy. The solutions came only about a hundred years ago. Today we're going to take a look at one of the most interesting of Zeno's paradoxes."

Nico went to the board and drew as he spoke. "Zeno asks us to consider a runner starting at a point S. He is going to his target T, which is 1 mile away. Now, to get to T he must first get to the midpoint between the starting point S and the target T. Call the midpoint M_1. It is a half-mile from T."

Start = S M_1 Target = T

So far, so good. I was getting interested. There was always the faint hope that I would be able to crack a problem even though it had shown itself to be extraordinarily difficult. Nico had reintroduced me to the pleasures of the mathematical hunt; I would be chasing a puzzle!

Nico, meanwhile, was busy drawing another picture. "To get from M_1 to T the runner must once again get to the midpoint between the two. Call this midpoint M_2; it is a quarter of a mile away from T."

I began to see where this was going. To get from M_2 to T, the runner would have to get to M_3, then M_4, and so on forever. Nico's next drawing confirmed this.

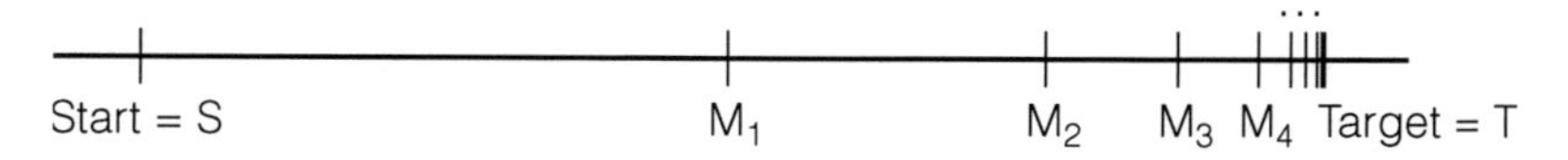

"Because I'm constrained by a chalk of finite thickness I have not drawn M_5, M_6, and all the other M_n out there. But Zeno argued that the runner would indeed have to pass through an infinity of these points," said Nico. I could tell Nico was getting excited—his voice was louder and his pacing more intense. "You see," he said, walking over to the blackboard, "no matter how close the runner gets to T he still has to cover half the remaining distance, then half of what's left, and then half of what's left yet again. Essentially he has to keep making runs between successive M_i's. First he runs between S and M_1, then between M_1 and M_2, then between M_2 and M_3, and so on. Let's call each such run an 'M-run'."

Nico went to the blackboard again and wrote:

1. The runner would have to make an infinite number of M-runs.
2. It is impossible for the runner to make an infinite number of M-runs.
3. Therefore, the runner will never get to the target.

"Historians can't really be sure how Zeno himself saw his paradox. He may have seen it as a logical conundrum, or he may have used his argument to conclude that all motion is an illusion," said Nico.

"That's crazy!" exclaimed Peter. "Motion is not an illusion!"

"I agree, it sounds utterly crazy." said Nico. "Clearly, motion is possible. Clearly, people move and cover distances. Clearly, a runner can cover a mile without getting trapped within a sequence of M-runs." Nico was pacing again. "But just as clearly, logic works in our world. If an apparently logical argument leads to an absurd result, then either logic does not always work, or the argument is flawed in some subtle way. I firmly believe that logic works. So there must be a subtle flaw in Zeno's argument. And I want us to find it."

Peter nodded. Considering his impatience with philosophical arguments of this nature I knew that Nico had gotten through to him. I myself was beguiled by Zeno's argument. It seemed extraordinarily simple, trapping one in its iron-clad logic, and it was hard to avoid hurtling toward its inevitable but absurd conclusion.

"Let's examine the argument in pieces," resumed Nico. "First, does anyone doubt that the runner would have to make an infinite number of M-runs?"

"I do," said PK. "Toward the end the M-runs become so small that the runner's body itself would span over all of the last few M-intervals and would cover the target."

"Valid point. Anyone care to shoot that down?" asked Nico.

Adin spoke up. "Nothing stops us from assuming that the runner is a dimensionless point that will have to travel a finite distance no matter how small the M-run."

"Exactly correct," said Nico. "I think Zeno was indeed correct in saying that an infinite number of M-runs would have to be completed for the runner to reach his target. But the next leg of his argument is even more interesting. He claims that it is impossible for the runner to complete an infinity of M-runs. What do we think of that?"

PK thought that there was no question Zeno was right. "If the runner has to cover a finite distance an infinite number of times he would never reach his target."

"That certainly seems correct. Anyone see a way out?"

The class was silent. If you kept adding a diminishing but finite quantity to itself forever, wouldn't you get a sum that grew forever?

Nico wanted us to think with specific numbers. "Let us say that the runner keeps a constant pace of running one mile in four minutes. How long will it take him to run a half-mile?" asked Nico.

This was simple. "Two minutes," someone said.

"And a 1/4 mile?"

"One minute."

"Right. 1/8 mile takes 1/2 a minute, and I'm sure you see the trend. Now if we sum up how long it will take the runner to complete each M-run we get an infinite series because there are an infinite number of M-runs." Nico wrote out the series on the blackboard.

$$\text{Time taken by runner} = 2 + 1 + 1/2 + 1/4 + 1/8 + 1/16 + \cdots.$$

"As I'm sure you've realized the 2 is for the first half-mile, the 1 is for the next 1/4 mile, and so on. The three dots denote the unending nature of this series. Now one objection Zeno might have had is that this series grows unboundedly large. Would he have been correct?"

Nico's question hung in the classroom. Calculators were pulled out, pens uncapped, and notebooks opened. The class was experimenting. I just stared at the unending sum. I recalled from the Bauji days that an infinite number of additions could yield a finite sum, but I couldn't immediately see how that was possible. After all, wouldn't one always have more terms to add?

Meanwhile, Peter had some interesting statistics to report. "If you add the first five terms, the sum is 3.875," he said. "If you add the first ten terms you get 3.996. Adding more terms gets you closer to 4, but the terms get smaller so quickly that they contribute almost nothing to the sum. It seems very probable to me you can never pass 4 no matter how many terms you add."

"Brilliant!" said Nico. He actually pumped his fist. "What do other people think of Peter's bold guess?" There was agreement in the room. Others had come up with similar calculations. "I see several nods," said Nico. "Peter, you are correct that the sum of the terms never exceeds 4. But in mathematics, unlike any other branch of human learning, you have to prove your case. It is not enough to simply state your result as a guess; an airtight justification is necessary, and without an airtight justification, nothing is resolved. This is not meant to discourage you," he said to the class at large. "In fact, I am telling you that Peter's guess is correct and that a series of infinite terms can indeed yield a finite sum. I will even tell you that the sum of the series is 4. So not only does the series not exceed 4, as Peter suggests, but I claim that it exactly adds up to 4. Once we understand *why* this is true we will have removed what may have been one of Zeno's concerns."

"I understand why," said a woman's voice from behind me. She appeared unaware that her declaration had come across as boastful for the emphasis had been on the "I." The entire class (even Adin) turned around to look at her. Her strong jawline was the first thing I noticed about her. Today, as she has entered her thirties, the rest of her face has softened a little, but the jawline remains as well defined as ever.

"You understand why? Have you studied infinite series in a Calculus class?" asked Nico.

"No, I just figured it out," she said seemingly unaffected by the scrutiny.

"What's your name?" asked Nico.

"Claire Stern."

"Well, Claire, come on down then, and show us," said Nico handing her a piece of chalk.

Without saying a word she wrote:

$$\text{Sum} = 2 + 1 + 1/2 + 1/4 + 1/8 + 1/16 + 1/32 + \ldots$$

$$-\ 1/2 * \text{Sum} = \quad 1 + 1/2 + 1/4 + 1/8 + 1/16 + 1/32 + \ldots$$

$$= 1/2 * \text{Sum} = 2 + 0 + 0 + 0 + 0 + 0 + 0 + \ldots.$$

$$\Rightarrow 1/2 * \text{Sum} = 2$$

$$\Rightarrow \text{Sum} = 4$$

She nonchalantly tossed the chalk back towards Nico, who was by then smiling broadly at her. "Claire, that is a beautiful argument! For the class, please explain what you've done here."

"We're trying to determine the sum of the series, so I put it on the left-hand side," she said. "I wrote out the infinite sum on the right, the way you did earlier. Next I multiplied both sides of the equation by 1/2 and shifted all the terms over by 1. I subtracted the second equation from the first, and all the terms got cancelled. The tails in the two equations are exactly the same and may be zeroed out upon subtraction. It left me with $1/2 \times \text{Sum}$ on the left-hand side, equaling 2 on the right-hand side, which implies that the Sum = 4."

"Bravo! Bravo! That is such an elegant argument, Claire. I love the way all the terms seem to cancel out. I'm proud of you," Nico said.

I could tell he was immensely pleased with what Claire had just demonstrated, and so it was a true surprise when I heard him announce, "Claire's proof is clever, but it is not correct," as she headed back to her seat. Nico waited until she was settled in and then spoke directly to her. "Claire, you applied the rule of finite mathematics to an infinite sum. In a finite equality you can multiply both sides of the equation by 1/2. But what does it mean to multiply all terms of an infinite sum by half? How do you do that?"

"I don't know what you mean," Claire told him.

"You cannot treat infinite series as finite ones. Strange things can happen. Let me give you an example," he said, going to the blackboard. "It involves rearranging the terms of an infinite series."

$$\begin{aligned} 0 &= (1-1)+(1-1)+(1-1)+(1-1)+(1-1)+\cdots \\ &= 1+(-1+1)+(-1+1)+(-1+1)+(-1+1)+\cdots \\ &= 1+0+0+0+0+\cdots = 1. \end{aligned}$$

"I've done nothing but rearrange the terms, which is perfectly legal in a finite sum, but I get an absurd result of 0 being equal to 1. Claire's argument leads to the right answer—unlike what I just showed you—but it yields the right answer without the right method. To fully understand and appreciate what it means for an infinite sum to converge to a point, we have to tackle the fascinating idea of limits. We'll get into that next week."

Nico's example was exactly on point. I saw that infinite sums are strange animals and may not always behave the way one might expect. For the first time in a long time, I felt the stirrings of interest; I wanted to figure out what was going on with these series.

Meanwhile, Nico was summarizing the discussion so far. "Zeno's argument said that it is impossible to make an infinite number of M-runs. One reason for his claim may have been the assumption that an infinite number of M-runs must add to an infinite distance. We've begun to see that this need not be the case, and we will complete our understanding in the next class."

Nico's passion for accuracy impelled him to leave us with a warning. "Zeno's paradoxes are one of the most discussed pieces in the history of philosophy. It is entirely possible that Zeno himself had some other reason for thinking that it is impossible to have an infinite number of M-runs. But all I expect you to take away from our discussion is that an infinite number of terms can and do converge to a finite quantity, which by itself is one of the great ideas of human history. The Greeks never fully understood this; actually, no one did until a hundred or so years ago, when the concept of limits began to be developed. Many mathematicians used heuristic methods to develop the mathematics of infinite series, but their methods—like Claire's fantastic argument—lacked rigor. As we saw, Claire's argument was almost correct, but we'll make it air-tight by taking a closer look at limits and infinite sums next week. Right now let's take a quick break. When we get back I'd like to talk about the treatment of infinity through medieval times. There are stories of great stupidity and great genius. Ten minutes."

• • •

Peter introduced me to Adin during the break. "This is my roommate Ravi," he told him, "and Ravi, this is Adin. Adin and I used to work out together. We used to race each other in the swimming pool every morning." Both Peter and Adin were tall, but whereas Peter was muscular, Adin was slender.

"Who won?" I asked.

"It was always pretty close," said Peter. "That's what made it fun."

"Peter would win in the shorter lengths, but I could usually take him if we swam more than five laps," Adin said.

I told Adin that I remembered seeing him play at the Coffee House. "You play the sax, don't you?"

He was obviously pleased that I had remembered. "Yeah, a little bit."

"So why is a musician-philosopher taking this math class?" I asked him.

Adin laughed. "Musician-philosopher! I'm afraid I'm neither. But I am interested in both—for very different reasons. This class actually connects up pretty well to several key ideas I'm interested in. Math has a great deal to say about philosophy."

"Really?" I was surprised. "Like what?"

"Well, Nico touched upon it in class. Like he said, mathematics requires proof, and proof confirms truth. I've always been interested in how one can be sure of something, and mathematics seems to provide the way to certain truth. Certainty is very important to me."

"What do you want to be certain about?" Peter asked.

"The purpose of life, for instance," he said, without missing a beat.

Peter laughed. He thought Adin was joking. But Adin's faced remained earnest and steady.

• • •

After the break, Nico drew a curious-looking drawing of two concentric circles on the blackboard.

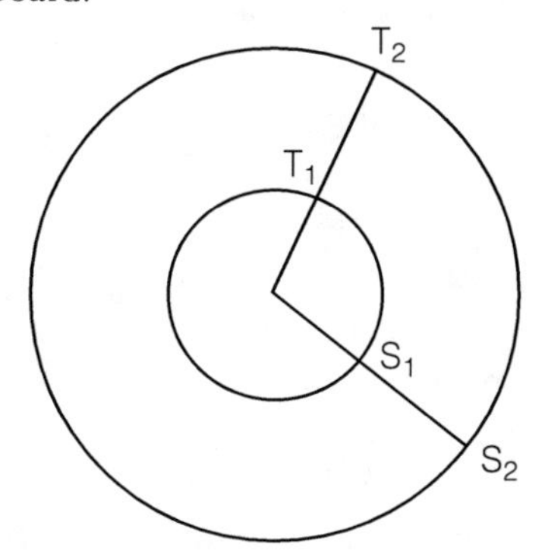

"Zeno's paradox was not the only problem that worried mathematicians through history," he said. "There was also this curious example of two concentric circles. Each circle has an infinity of points on its circumference, but since the inner circle is smaller, one would think that it contains fewer points than the larger outer circle. But if we draw a radius to the outer circle, we can see that each time the radius touches a point on the outer circle, it also touches a point on the inner circle. So the sweeping radius sets up a correspondence between points on the circumference of circles of different sizes. In our picture S_1 corresponds to S_2 and T_1 to T_2. This correspondence seems to suggest that the two circles have the same number of points, even though one is bigger. How can that be?"

We sat there looking at the circles. There was no doubt in my mind that for every point on the larger circle there was a point on the smaller circle. But this defied common sense! Surely the larger circle had more points. Meanwhile, Nico amused himself by making slurping sounds in his coffee cup and said nothing further.

Finally, it was Adin who thought he had found a way out: "I agree this looks very strange, but unlike Zeno's problem there is no logical paradox here. Zeno's suggested that there could be no motion, which we know to be false. All this problem is saying is that the two circles have the same number of points, which is strange, but in my opinion there is no contradiction here."

"That's pretty good, Adin," said Nico. "You're right, it's strange, but the strangeness might be just because many of us don't have strong intuitions about infinite sets." I saw Adin's point, yet was not fully satisfied. He was right—there was no apparent logical contradiction—but I was not comfortable with the result, even though the evidence was in front of me.

Apparently, my uneasiness had been shared by other mathematicians through history, who, according to Nico, concluded from such results that infinity was a slippery, even dangerous concept. "Until modern times, most of what was written on infinity after the Greeks was more theological than mathematical," Nico said. "Medieval mathematicians saw infinity as an awe-inspiring and sometimes a fear-inspiring idea. 'Only God is infinite' was their conclusion; everything else is limited. An Italian thinker, Giordano Bruno, was tortured for nine years in part because he refused to retract his idea that the universe was infinite and extended forever. Bruno believed that reason and philosophy are superior to faith, and to knowledge founded on faith. He refused to accept the finiteness of the universe merely because the Church decreed that only God could be truly infinite. At his trial, which ended in 1600, he was as defiant as ever. Upon hearing his death sentence, he responded, 'Perhaps your fear in passing judgment on me is greater than mine in receiving it'. He was then gagged and burned alive."

A shocked "Oh God!" escaped out of Claire. It came out louder than she had intended.

"I know; sometimes life is terrible," said Nico nodding in her direction. "Bruno was killed by people who valued power over truth. Unforgivable." It was the only time in the entire semester that I saw Nico look angry. His usual amused half-smile was replaced by a downward scowl that deepened the lines running down along his nose. His eyebrows crept together, and his large, usually clear forehead became riven with wavy furrows. It seemed to me that ideas were as important to Nico Aliprantis as they had been to Giordano Bruno.

After a while he continued: "The one shining exception to mediocre thinking about infinity before Cantor was Galileo."

"The telescope guy?" asked PK.

"The very same. Galileo, who invented the telescope and first saw the moons of Jupiter and the rings of Saturn, also had an insightful idea about infinity."

He turned to the blackboard and wrote:

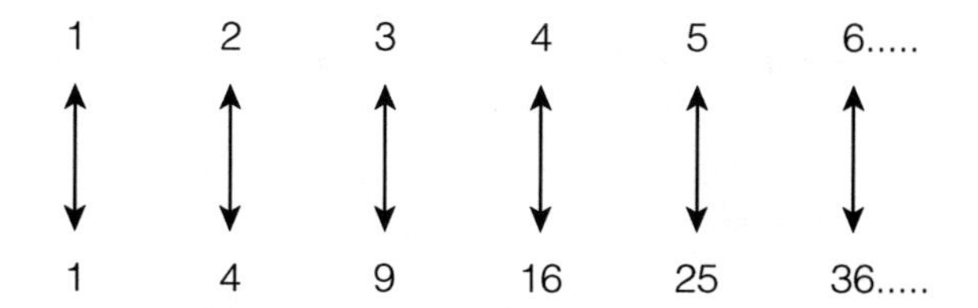

"Galileo observed that you could put a number and its squares in a one-to-one correspondence. There is an infinity of numbers and an infinity of square numbers. This seems to show that there are as many numbers as there are squares, but at the same time there are a lot of numbers that are not squares." Nico was pacing now, his coffee mug forgotten. "Many people before Galileo had observed this correspondence and they had deemed it a paradox. 'How can a part be equal to a whole?' they asked. Galileo's insight was this: *he realized that you cannot apply the laws of finite mathematics to infinite sets.* Claire tried to do that earlier, and even though her method was elegant, I had to stop her. Galileo said that for infinite sets it is possible for a part to equal the whole; that this was no paradox, only a property of infinity. With this simple conclusion, Galileo introduced the modern age of infinity. And that is as far as we are going to go today."

It had been a perfect class. Nico had been clear, engaging, and stimulating. Half of me felt like standing up and giving him a hand. Of course I did no such thing. But now, many years later, from my perch in adulthood, I wish I had.

"Before you go," said Nico, "I'd like you to spend 10 minutes thinking about a simple question related to the concentric circle problem we discussed earlier. He drew two straight lines, one longer than the other.

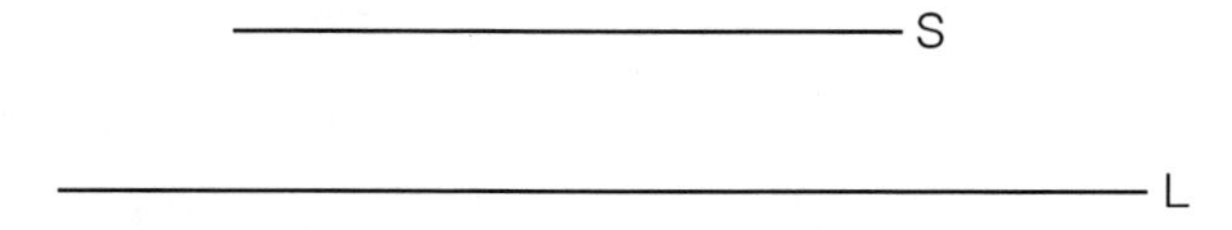

"Let's label the shorter line S and the longer line L. Here's my question: Does L have more points than S, or do they have the same number of points? Please prove your answer, write it down, and drop it off before you go." With that Nico stopped talking. He found a seat, something to read, and an old cigar to chew on.

My first instinct was that the number of points had to be the same. If two circles of different lengths could have the same number of points, then surely these two lines would as well. But try as I might I couldn't find a mapping that led to a one-to-one correspondence. I was missing the equivalent of the center of the circle.

And then I saw it. Aha!

I quickly wrote my answer: Draw 2 lines connecting the pair of endpoints. Extend these lines to intersect at P.

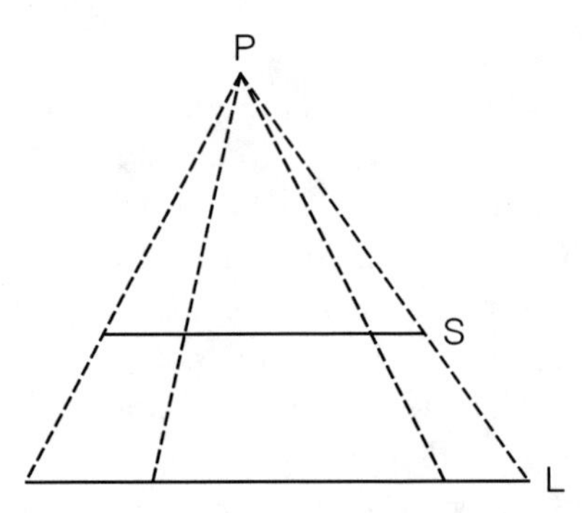

Now the lines that pass through P and connect S and L establish a one-to-one correspondence. Therefore, S and L have the same number of points. I handed my answer to Nico on the way out. He looked at it, nodded, and then grinned. But I hadn't been the first to finish. Claire Stern was finished almost as soon as Nico had presented the problem. By the time I got out, she was gone.

When Adin and Peter didn't come out of the classroom ten minutes after everyone else had left, I went in to see what was going on. I found them embroiled in a heated discussion.

"Ravi, take a look at this," said Peter. "Adin says this is not correct." Peter showed me his drawing:

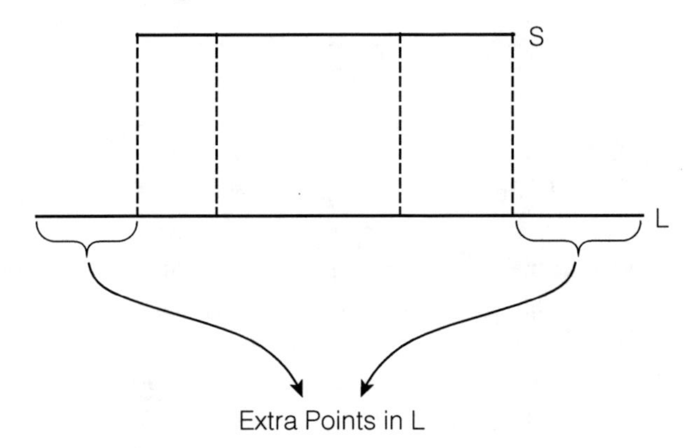

"This shows that there are more points in L. I've mapped everything in S onto L, and L still has points left over," said Peter pointing to the parts of the longer line L that extended out from the section directly under S.

"Peter, I'm not saying that your mapping is incorrect. I'm saying that it doesn't prove anything," said Adin. "We are here required to show one of two things: either that there exists some one-to-one correspondence between the two lines, in which case they are of the same size, or that no such correspondence can exist, in which case one is bigger than the other. Finding a correspondence that is not one-to-one doesn't really show anything. There could still be another correspondence that is, in fact, one-to-one."

"I don't get it," said Peter, looking towards me.

I saw what Adin was getting at. He was right, but was not explaining himself too well. Peter needed an example.

"Peter, you remember the correspondence where Galileo matched each number to its square?" I asked.

"Yeah," he said, unsure of where I was taking him.

"Suppose I match each square with itself," I said, asking for his pen. I wrote out an alternative correspondence:

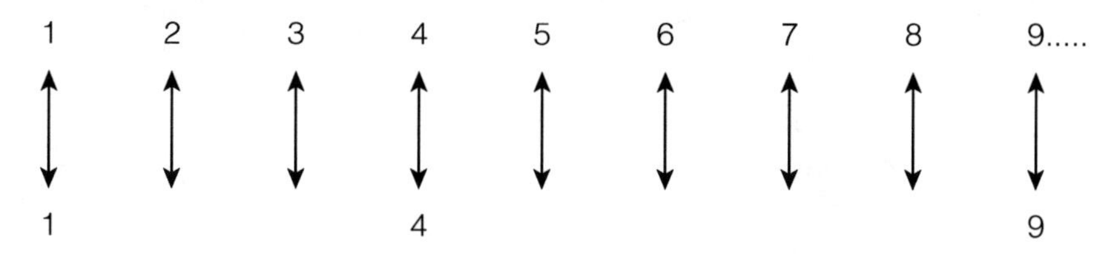

"See, Peter, here all the squares are matched up, and there are all these numbers left over which makes it seem like there are fewer numbers in the bottom row. But this correspondence does not show anything, because we know another correspondence exists that matches the collections in a one-to-one manner."

Peter got it. He hit himself on the back of his head. "Duh! I should have seen that," he said.

• • •

Later that day I saw Claire. She was sitting in the main quad, her back very straight, with a notebook in her lap and a tapping pencil in her left hand. As I approached her and could make out the drawings in her notebook, I saw that she was still trying to map Nico's shorter segment into the longer one.

“You were the first one to be done with that; how come you’re still working on it?” I asked her, quashing the impulse to play it safe and walk on without a word. She looked up reluctantly, a little annoyed, I felt, to have had her thought stream intruded upon. The sun reflected off the lenses of her glasses and I couldn’t see her eyes.

“I wasn’t too happy with my proof,” she said. “I’m looking for a more direct approach.”

She didn’t invite me to sit; she didn’t ask my name and she didn’t inquire if I had solved the problem. Instead, she went back to her notebook, leaving me standing there, feeling like a schmuck.

I stubbornly refused to leave. Pulling out a sheet of paper of my own (not daring to ask her for one) I drew out my solution that joined the two endpoints and extended the two lines until they met. Brashly, and without a word, I placed my drawing on top of her notebook.

She looked at my drawing for about a minute and then nodded. “That’s good,” she said. And then she shifted to the left, making room for me. She pointed to her notebook. “My idea was that any line of a given length can be transformed into a circle with a circumference of that exact length. So with the two lines you can draw two circles with unequal circumferences, and we’ve already shown that all circles of any circumference have the same number of points.”

It was an indirect proof, and in its way it was quite efficient. It used the result that we’d already proven in class.

“That’s clever,” I told her.

She shook her head. “No, it’s not as direct as what you did. I think yours is what Nico was looking for.” She shut her notebook, preparing to leave. Impetuously and impulsively and risking rejection I asked her if she wanted to grab lunch at Tressider. Much to my surprise she said yes.

On our way to lunch we passed by Nico’s office. The door was open and he saw us walk by. “Hey, you guys, can you come in here for a minute?” Nico was sitting with his legs on his desk looking at everyone’s responses to his question. “You two were the only ones who got the answer right, although your approaches were quite different.”

He asked us how we liked the class.

“It was stimulating,” said Claire.

I was more effusive in my praise. “That was one of the best classes I’ve ever had. I don’t think they can get any better, only different.”

Nico smiled. “So it meets the standards set by your grandfather?”

Once again I was surprised at his memory. “Absolutely,” I told him.

"I meant to ask you—your grandfather was a professional mathematician?"

"Yeah. He never formally got his Ph.D., but he published a lot of papers."

"What was his field?" asked Nico.

"He published in many different areas, but his concentration was in algebraic number theory."

"You're kidding!" said Nico. "That was the subject of my dissertation. What was his name?"

"Vijay Sahni."

"Vijay Sahni . . . . I know that name. Wait a minute, did he do work on elliptic fields?"

I didn't know. "He might have," I said.

"I do know that name. In fact I think your grandfather may have written a paper that was a particular favorite of mine. Hold on a minute, let me find it." Nico's bookcase must have been more organized than it looked, for a few minutes later he had found what he was looking for. He fished out an old book titled *Classic Papers in Algebraic Number Theory*. Most of the book's pages were heavily underlined.

"This book used to be a great favorite of mine. It was published in 1961 and not many people noticed it when it came out, but I must have spent endless hours with it," he said quickly turning the pages. Then he found it. "There it is! 'Elliptic Curves Over Function Fields' by Vijay Sahni. Is that how he spelled his name?" Nico asked, pointing to the open page.

It was his name.

"Look at the footnote," said Claire, looking at the page from behind me.

I did. There it was, in smudgy black and white. The handful of words after which things would never be the same: *Note from the Editors: Mr. Vijay Sahni informs us that the key ideas contained in this paper were formulated while he was serving a prison sentence in Morisette, New Jersey, in 1919.*

I did know that Bauji had come to America towards the end of the Great War. But Bauji in prison? There must be some mistake. Yet, the paper was in his field. And the date mentioned in the footnote made sense—surely there could not have been more than one Vijay Sahni in the United States in 1919. Could it be that he had been imprisoned? But for what crime?

And what in the world was he doing in Morisette, New Jersey?

CHAPTER

1 2 3 4 5 6 7 8

"Hello?"

"Ma, it's Ravi."

"Ravi? It's really early in the morning here. Are you okay?"

"I'm fine. There is nothing to worry about. But I wanted to talk to you about something important."

I could hear her sitting up on the bed. "What's the matter? Are you okay?" she asked.

"I'm fine. Ma, you remember that Bauji was in America in 1919?"

"Yes, he'd gone there from England. I think some professor had invited him because he proved some theorem. Why?"

"Where in America was he, do you know?"

"He first went to New York and then to a small university town, whose name I'm forgetting. . . . What is all this about, Ravi?"

"I'll tell you in a second. Do you know which state this town was in?"

"It was near New York, because he used to go there some weekends."

That just about settled it. "Does New Jersey ring a bell?"

"Yes, it does!" she said excitedly. "It was Morisette, New Jersey. Now I remember the name. He corresponded with someone from there. We used to get letters postmarked from Morisette when I was a child. But what's going on Ravi? Why do you ask?"

"I just found out that Bauji was in a prison in Morisette in 1919. It completely amazed me."

"Prison?" So she didn't know either.

"Yes, I saw an article by him in a book. The editors of the book said that Vijay Sahni had formulated the ideas contained in the article while he was imprisoned in Morisette, New Jersey."

She stayed silent for almost a minute, and I let her be. When she spoke it was in a quiet, faraway tone. "Ravi, he never explicitly said anything about being in prison but a few things did slip out over the years. Once, when I asked him what America was like he said that it was beautiful and free, and that even the prisons there were quite comfortable. When I asked him how he knew, he looked at me as if he was wondering if he should tell me . . ." Her voice trailed off, thinking back to a conversation that was probably over 40 years old.

"I remember his look like it was yesterday, Ravi. I could see him struggle with the decision about whether or not he should tell me something, and that's the first time this idea began to develop in my head that maybe something happened in America. And then every few years there would be another snippet that would make me surer of my hypothesis."

Hypothesis indeed. She was, after all, a mathematician's daughter, and was quite comfortable with the lingo.

"Like what?" I asked

"Well, I remember he got completely fascinated by a book about escapes from prison. He used to read it and reread it." Suddenly I remembered that the only movie I remembered him seeing more than once was *Escape from Alcatraz*.

"And all his life, Ravi, he retained a great interest in American politics, especially the idea of separation of church and state. In fact, he even wrote about it in the Indian context. I always thought that it had something to do with his time in America. But what that could have to do with prison, I don't know. At some point I just decided not to worry about it."

"Ma how long was he in America?"

"I'm not sure, Ravi. All this happened more than 20 years before I was even born."

I hadn't realized that, but of course she was right. Ma was born in 1940. Bauji had her when he was 45. So in 1919 Bauji was 24 years old, and my mother was over two decades from being born. "So, do you have any idea why he may have gotten arrested?"

"Sweetie, I don't know. And you know, it was so long ago, I'm not even sure it matters. If something happened he probably had good reasons for not wanting to tell us. Why not just let it be?"

"Because it was a part of him. I feel like I loved him completely and now it turns out I didn't even know a big part of him."

"I know. I know. But Ravi, maybe the whole thing wasn't that important. Things happen in people's lives. They take care of them and move on. Just because Bauji never said anything about this one incident—that may or may not have happened—it doesn't mean that it was a crucial part of his soul, or that you didn't know the real him."

I could see her point of view, but I knew that I wanted to know what had happened, and this wasn't going anywhere. "You're right. Maybe it doesn't matter and maybe it'll be very hard to find anything out in the first place."

"I know you Ravi," she said after a pause. "I know you're going to do everything to find out what happened. Whatever it is, do not let it spoil your memories of him."

Fair enough. "I'll try not to do that."

"So, how is everything else? You must be signed up for five economics courses this semester?" Her voice was firmer, and her words less hesitant. We had switched from the past to the future.

"Actually no, I've signed up for four."

"Four? I thought you had to do five to get your major in economics."

I hadn't really planned how I would break the news. Now there was no time for any spinning. Truth time. "I found another class I really wanted to take."

"But then you can't graduate in the spring. You yourself told me you needed to take five courses in economics. What is this other class you're taking?"

"It's a math class. I really like the instructor. I'm really enjoying it."

I shouldn't have added that last bit. "Enjoying it? You think you are there to be entertained? How do you expect to graduate now?"

"I'll take six classes next semester or maybe I'll take an extra class in the summer."

"And who's going to pay for that? You know the money Bauji left is already used up."

"I know. I'll figure something out. Look Ma, I just found out this thing about Bauji and you're giving me a hard time on top of that."

She didn't soften. "Ravi, you need to get your major requirements done, sign up for some interviews, get a job, and make enough money to justify all this expenditure on your education. This whole thing with Bauji's imprisonment is a distraction. Do not spend time on it. You need to be earning by this time next year. Our family is counting on it. You know what the money situation is."

"I will be earning, Ma. I'll be signing up for some interviews this week."

But she hadn't heard me. "This thing about you having to take a summer class to graduate is a new worry. I wish you hadn't done it. I don't know where the money is going to come from."

"I'll figure something out."

I heard her sigh. The conversation had run its course. "Okay, then," she said in closing.

"Okay, Ma. Sorry I woke you up."

Peter walked into the living room just as I was finishing the phone call. "Everything okay?" he asked.

"Not really." I told him about Bauji's imprisonment and about the financial impact of having to do an extra course in the summer.

"Dude, your grandfather's thing—that was 80-something years ago! It doesn't matter now."

Peter did not believe in time. He had told me more than once that what happened in the distant past (by which he usually meant anything before last year) is irrelevant to what will happen tomorrow, and it's tomorrow that matters.

"Your grandfather was probably arrested for some little thing and he did some math when he was in prison, but so what? How could it possibly impact your life now?"

I could have told him that I found myself to be much like Bauji was: I have his coloring and his nose; I, too, like trumpet solos and overflowing bookcases; I am annoyed by long phone conversations and amusement park rides; I even have his angular handwriting, his taste for mangoes, and his abject lack of drawing ability. How could I not care what had happened to him? I am him.

But I didn't have all this clearly formulated then, so I said nothing. I was just aware of an imperative to find out what had happened to Bauji and couldn't really say why. Without intending to, I shrugged.

Peter mistook my shrug for agreement and switched topics. "Now, about the other thing, I thought you weren't going to sign up for Nico's class. I was surprised you did."

"Peter, I had to. It's been so long since there was a class I was excited about."

Peter nodded. He had heard this theme from me before. I'm sure he didn't fully understand the exact cause of my dissatisfaction with most of my classes, but nonetheless he had come to accept it. I saw him squint his eyes, a sure sign that something was cooking inside. He could make quick decisions on incomplete information more effectively than anyone I have ever met, before or since. "Listen Ravi. Come summer, if you still need to take the class and if you still need the money, then I'll loan it to you. I have the signing bonus coming in next month, so I'll have the cash."

It was a generous offer, one I was in no position to refuse. He interrupted me before I could thank him (he hated even the hint of sentimentality). "I've got to get to the gym."

• • •

The next morning as I was riding my bike back to my room in Escondido Village I heard Claire call after me. She was in the garden of her ground floor apartment with shears in one hand and some stray branches in the other.

"Don't you need gardening gloves?" I asked.

She laughed. "Real gardeners avoid gloves whenever we can. It's nice to feel the plants with one's bare hands," she said.

So she was a gardener—a *real* gardener at that—and her skills were apparent. At that time I couldn't say exactly why her garden looked good. It was not especially neat or big or colorful, but it was a fine space that was pleasing to look at. She would teach me later that the trick is to pick plant combinations that fit well together, and also fit in the space they were going to occupy.

"There's a bug crawling up your shoulder," I told her.

She looked at it from the corner of her eyes. "Ah yes, he's an Assassin bug, a friendly. They eat the pests that threaten my plants," she said, gently flicking him off.

I was impressed that she could identify the insect without pause. "Wow, Claire, you seem really into this. I, on the other hand, can kill any plant I buy within a week," I told her.

"That can't be true!" she laughed again. "All it takes is patience and stick-to-itiveness." Then she looked into my eyes and added, "Like I think you have in mathematics."

I didn't know quite what I would say to that and my hesitation must have shown, for she quickly changed the topic. "I've been thinking about your grandfather."

Her view was quite different from Peter's. "You *must* find out what happened. I've been thinking we should go visit my mom. She'll know how to find out what happened."

"Your mom?"

"Yep. She's an Information Specialist at the Graduate Library. Let me get my bike and we can ride over there."

Carol Stern turned out not to look much like her daughter. Her hair was curly and long where Claire's was straight and cut to above her shoulders, her lips were full and curvy—probably twice the surface area of her daughter's—and she didn't have Claire's penchant for wearing black. Only their eyes were the same: big, clear, and curious. Claire spotted her as soon as we got to the library. She was walking towards the chemistry shelves pointing something out to a student who was trying to keep up with her long strides. Claire waved and caught her eye. She smiled broadly and mouthed "five minutes" from across the room.

While waiting for her I decided to try an Internet search for Bauji. It was 1990, just a few years before the World Wide Web spawned, but even without Google there were reasonably effective ways to find things. I selected some likely looking databases and typed in "Vijay Sahni."

There were about 150 search results. The first seven entries were about someone who appeared to have written extensively on earthquake retrofitting. The next few were from a plastic surgeon who promised a "more confident you." Then there was a student at the University of Chicago, a software engineer in Santa Clara, and an immigration lawyer in Montreal.

Vijay Sahni after Vijay Sahni, but no Bauji.

Claire introduced me as "my friend Ravi" to her mother. Carol looked at me carefully, making me think that Claire probably didn't use the word "friend" loosely. Carol's handshake was firmer than I would have guessed.

Claire told her mother about the class, Nico's old book, Bauji's article, and the footnote. "Ravi knew nothing about this imprisonment and he really loved his grandfather, so we need to find out what happened here."

I had not told Claire how I felt about Bauji; she just knew. Carol nodded, seeming to understand the imperative to find out. "Come on, let's go to my office."

Her office seemed to have as many books as were in the library itself. Every available surface was piled five to six high with books of every size. The three large bookcases on the walls had long since been overwhelmed. There were books of every size, shape, and topic, though there seemed to be a few themes interwoven as well. I detected several books on animal rights, Jewish history, and Dutch art.

"Sit, sit," she said, moving a pile of books from the chair to the floor. "Tell me about your grandfather."

I told her the biographical facts as I knew them: he was born in the Punjab, in Northern India, in 1895. He died in 1980, alert and communicative to his last day. His parents were well-to-do farmers and landowners, so he had access to good schools even though the British were in control of India when he was growing up. Early on, he showed a flair for the sciences, and particularly for mathematics. He wrote two notable mathematics papers when he was 16. G. H. Hardy, a famous English mathematician, saw his papers and invited him to Cambridge.

"Hardy was the guy who discovered Ramanujan, wasn't he?" Carol asked.

Ramanujan was an Indian mathematician, a genius of the highest order. He was rescued from obscurity by Hardy, who saw sparks of brilliance in a letter he received from Ramanujan, at that time an anonymous clerk in Madras. Two other prominent mathematicians had rejected Ramanujan's letters as ravings from a crackpot. But not Hardy. Within an afternoon he knew that the letter was the work of an extraordinary mind. Soon enough, Ramanujan was on a boat to England.

"Actually, my grandfather knew Ramanujan," I told Carol. "He used to say that Ramanujan's mathematics was so beautiful it almost made him believe in God."

"He was an atheist?"

"In a way," I replied. "It was almost as if he wanted to trust in God, but couldn't quite allow himself to do it."

Carol nodded slowly.

"Anyway, he spent a year or so in England, and then in late 1918 or possibly in early 1919 he came to America, again at the invitation of a mathematician."

"Where in America?" asked Carol.

"Apparently to a place called Morisette, New Jersey."

"Is there a university there? Because if there is, I'm sure they'll have records."

"There must have been, but I don't know if it still exists. I've never heard of a Morisette University."

"We'll find out. Go ahead with your story," said Carol. Her interest made me feel that I was in good hands.

"My family isn't quite sure how long he spent in the United States," I told her. "But we know that he was back in India by 1924, because that's when he bought the family home in Delhi. We know almost nothing about his life from 1919 to 1924, and I had little reason to find out. Then yesterday, when I found this." I handed her a copy of the page that Nico had shown me. "Look at the footnote."

Carol read it, and reread it. The she looked at me, and read it one more time. "You didn't know he was in prison?"

“Never had the slightest inkling.”

“And you want to find out why he got arrested?”

“I want to find out everything. Why he got arrested, if he got any kind of a trial, how long he stayed in prison, and how he got out. Everything.”

“Right,” said Carol, sitting up straighter in her chair. “Who wrote this book that the footnote came from?”

“It is an old text, published in 1961. The author has long since passed on.” I had checked last night.

“Any idea how he could have known your grandfather?”

“He was from Cambridge, and was around Bauji’s age. I would bet they knew each other.”

“Bauji?”

“Sorry, that’s just what I called my grandfather. It’s the Hindi word for grandfather.

“I see. Okay, let’s start by looking for Morisette, New Jersey and any universities that may have existed there.” She pushed up her glasses, which had slipped down her nose, and swiveled around to face her computer. Her long skinny pianist’s fingers started playing the keyboard. She took notes as she scrolled through articles. In about five minutes, she found out quite a bit.

“Okay, Ravi,” she said, turning back to face me. “Morisette was essentially a paper-mill town whose population peaked at about 200,000 around 1860. In 1880 the family that owned the paper mill bequeathed money to build a university in the town so that their best students didn’t have to leave the state and go to New York City, which they perceived to operate on ‘questionable morals’.”

She had drawn the quotes sign in the air when she said “questionable morals.”

“Anyway, they wanted the university to be chiefly dedicated to engineering and the sciences, for they did not see great benefits from the humanities. They did, however, have a large divinity department, and to go with that a large church at the center of campus. The stated purpose of the university founder, a Mr. Gerald Westin, was to conduct research to develop techniques and products that would benefit all people.”

“So what happened to this school?”

“There are very few records, at least on this first search. It appears to have lasted until 1942, and then there’s this brief note in the Princeton paper about Morisette University shutting its doors.”

“That’s probably when every college-age male enlisted for the army,” said Claire from behind me.

"Good point. Also, it looks like the paper mill had closed down in 1937, and Morisette became something of a ghost town. Today its population is listed at 300."

"What closed the paper mill down, do you know?" I asked.

"Yes I do, actually. It says here that there was a major fire in the town, which burnt down the mill as well as several city buildings."

Maybe Bauji had somehow been responsible for the fire. But wait, he couldn't have been. He was long gone by that time.

"So what's the logical place to look for an account of my grandfather's arrest?" I asked.

"Well, several things come to mind. We could look for police records, court papers, or maybe . . ." I could see she had an idea. "Newspapers! A town that size would have had a newspaper." She rolled her eyeballs upward in a "why didn't I think of that earlier?" gesture.

I, however, did not quite see where she was headed. "But would we be able to get those newspapers?"

"Typically someone has them on microfiche. But first let's see if the town had a newspaper at all."

Carol turned back to the computer. "I do have this database of every newspaper ever printed in the United States." It took her only a few seconds. "Here it is! It was called *The Morisette Chronicle*." After a few more mouse clicks she began to read aloud. "*The Morisette Chronicle,* founded in 1872, was a community newspaper serving the town of Morisette, New Jersey. It was primarily focused on covering local news and statewide sporting events. The daily continued operations until 1935, at which point it bowed out to competition from the local editions of larger newspapers from Jersey City and New York City. All editions after 1888 are available in the New York Public Library."

"Wow," I said, "librarians keep every edition from every newspaper that ever existed?"

"We try to do that, but we fall way short. In this case, though, it looks like we got lucky."

"Thank you, Carol. This could be exactly what I am looking for."

"Don't thank me yet. We haven't got anything in our hands. But I'm hopeful. Tell you what, I've got to get to a meeting, but I'll call New York tomorrow, and I'll let you know what they say. If they have this stuff scanned in, they could just e-mail the files over. But most likely it's still sitting on microfiches. In that case we'll have to ask them to mail them over."

"How long would that take?" I asked, afraid that it might be too long.

"Could be several weeks, but they usually give you an overnight option."

I assured her that I would willingly pay whatever extra amount they required. I just wanted to know as soon as possible. "You'll know later today or tomorrow," she said, nodding her head reassuringly.

Carol was true to her word. There was a message on the machine at home. "Ravi, I called the New York Public Library. They will send us a copy of the microfiches. They can only send about 6 months worth of newspapers at a time, so I asked them for the first 6 months of 1919. I figured that was the place to start. I've asked them to overnight the package. It'll be here Friday morning, around 10:30. Bye."

• • •

Peter's parents threw him a party to celebrate his job offer from Morgan Stanley. They invited about a hundred friends and relatives to their home in Sausalito, a wealthy suburb of San Francisco.

Claire, Adin and I had driven up from Stanford together, and after a warm welcome Peter had invited us to hang by the pool in the backyard of his parent's quietly elegant home. The pool overlooked the Pacific, and if you sat low enough on the deck chairs it looked almost as if the two bodies of water gently merged into one another.

"This is the way to live," I told Adin.

Adin nodded. "Money is a good thing," he said. "In fact, the Sophists said it is the only thing."

I had heard the term "Sophist" used in a vaguely derogatory manner, but didn't really know who they were or what they stood for.

"The Sophists," explained Adin, "were a group of philosophers in Athens who pre-date Socrates. I first read about them in a philosophy book my mother gave me for my 16th birthday, and I was instantly hooked. Their ideas can be disheartening, but they're unflinchingly honest. They talked about things that I cared about then."

"What was their main idea?"

"They basically said that there are no absolute standards or values and that each person is the standard of his personal truth." Here was Adin's certainty theme again. It seemed to me that he was seeking absolute truth—a set of transcendent values that were true in every possible universe independent of the existence of human beings, or any beings at all.

Adin continued: "Protagoras was one of the leaders of the Sophist movement. He said that man is the measure of all things, meaning that whether something is considered true or false depends on who is doing the considering. He came to this conclusion because he traveled widely and saw so many different customs, social norms, and forms of government. Each society thought that their way was the natural way, the absolute way, and the certain way, but Protagoras and the Sophists realized that one way was not anymore certain or real than any other way. They went on to generalize that every human idea is relative to the circumstances surrounding the originator of the idea, and that true knowledge is unattainable. Therefore, it was best not to seek what one cannot find. Instead, they decided to focus on immediate things, like becoming wealthy and powerful in society. Money, they said, was a tangible good worth pursuing."

Sitting on that magnificent deck overlooking the Pacific, with the evening sun angling into our eyes, it was hard to argue with the Sophist conclusion. But clearly there was something other than money. What about knowledge? And love?

I asked Adin, and he shook his head. "What bothers me is not their conclusion that money is worth pursuing. It is worth pursuing, but as you say, so are many other things. What really bothers me is the fact that their statements suggest that all truth is relative, which to me, at least, means there is nothing to be absolutely certain about."

"But Adin, there is certain truth everywhere," Claire said. "We're sitting here on deck chairs at the edge of the continent, the sun is shining, and you're wearing a baseball hat. These are all truths, no?"

Adin shook his head. "Actually, Claire, there are people who would even deny you those things. Philosophers like Descartes have argued that there is no way of knowing that the deck chair actually exists. He said that while we have sense perception of the chair, it may not be there at all, or it may be a part of an elaborate dream."

He thought about what he had said and then distastefully crinkled his nose. "While technically irrefutable, these ideas are not interesting to me. I don't believe they're useful. I'm more interested in pursuing certainty about ideas," he said.

"Like what to do with your life?" I remembered him saying that last week.

"Absolutely."

"Adin, you can never be certain about that," I told him, from my years of considering the same question.

"Why not?"

I looked to see if he was joking, but once again he wasn't. And it wasn't a rhetorical question either; he wanted an answer.

"I don't know any idea that you can be absolutely certain about," I said, digging deep to give words to what I really felt. "Life is too . . . complicated. Truth seems to stem from personal circumstances," was what I came up with.

"But people have certainty about many things," he countered. "I know people who are certain that God exists. And for them this is an absolute truth—independent of all context. Others are certain that human life has meaning and purpose. They genuinely entertain no doubts; they have certainty."

"Adin, are you certain about any of these things?" I asked.

He shook his head. "I only have opinions, but no certainty. But I hope there's something one can be absolutely sure of."

"Really? You don't have certainty about ideas either?" I asked.

Adin shook his head and bit into some of the ice from his drink. "Perhaps no one has a thought he is absolutely sure of."

"I do," said Claire from behind us. She paused, enjoying the drama of her announcement.

"Well, what is it?"

"That there is no largest prime number."

This was completely unexpected. Adin and I had expected something philosophical, not mathematical.

"Claire, if you are allowing math and science, there is plenty to be certain about," I told her.

Adin disagreed with me. "Like what?" he asked, shaking his head.

"Like if I drop this glass it will fall and shatter."

"You don't know that," he said. "You believe it because of repeated observation. Observation cannot yield certainty. Just because you've always seen grey pigeons does not mean that one day you won't run into a white one."

He had a point. "Okay, how about 2 + 2 = 4? You can be certain about that, can't you?"

"Yes you can," said Adin, "but that's mostly truth by definition. You define the number 2 and the operation of addition and you're pretty much there. It's like saying that the Empire State Building is a building. I'm talking about ideas that need thought—questions arising naturally that we can ponder over in our minds."

"Such as the meaning of life?"

"Yes," he said matter of factly, "or maybe this prime number thing," he said looking at Claire, wanting to hear more.

"What exactly is this prime number thing you are talking about?" I asked her.

"You know what a prime number is?"

"Sure," I said. "Any number greater than 1 that is not divisible by any number other than 1 or itself. So 2 is a prime, 3 is a prime and 5 is a prime, but 4 is not a prime since 2 divides 4, and 6 is not a prime since 2 × 3 = 6."

"Yes, that's exactly it," said Claire. "What's interesting is that every number is uniquely decomposable into primes. For example, 45 = 3 × 3 × 5, and it's not decomposable into prime factors in any other way. It's in this sense that primes are the building blocks of mathematics."

Her example was well chosen. 45 = 9 × 5, but if you decomposed 9 you still ended up with 3 × 3 × 5.

She continued: "Anyway, since primes are the building blocks of all numbers, people have studied them very carefully. I, myself, am completely smitten by them." As always, she spoke with her hands. For "completely smitten" she brought her hands together, palms down, fingers interlocked.

Suddenly I remembered Bauji's note to himself to show me something about primes. But we had never gotten to them, and now Claire had taken up the thread.

"What fascinates me about primes," she said, "is their hidden order amidst their apparent disorder. On the surface they seem to follow no pattern at all. Yet everywhere there are tantalizing glimpses of hidden order."

"Like what?" I asked.

"There are so many things." Claire had a "where do I begin?" expression. "It's like asking what's beautiful about music," she said, looking at Adin. She was sounding a little like Nico.

"You said you're certain about something to do with prime numbers?" Adin asked, getting her back to his theme.

"Yes, I'm certain there can be no largest prime." She sat up straighter in her chair. "See, here's the thing. The density of primes keeps decreasing; in the first 10 numbers there are 4 primes: 2, 3, 5, and 7. That's 40%. In the first 100 numbers the prime density is down to 25%."

Claire fished in her bag to retrieve her ever-present notebook. "A couple weekends ago I ran a little computer program to figure out the percentages for larger numbers," she continued. "In the first 1000 numbers the prime density is down to 16.8%. If you go all the way up to a billion numbers, the prime density is only 5.08%," she said, consulting her notes.

Her eyes were shining. Something about prime numbers must have really grabbed her. "So the density of primes keeps going down?" I asked.

"Exactly," she said. "In fact, if you go high enough, the density of primes is nearly zero!"

Claire meant that primes got sparser and sparser as one got to the really large numbers. The next logical question was if they ever became extinct. In other words, was there a point after which every single number was non-prime? I asked Claire this and she nodded her head. "That's the right question to ask."

"My instincts," I told her, "rebel against this idea of there being a largest prime. I think the primes must get rarer and rarer, but they never vanish."

Once again she laughed. "Your instincts happen to be correct."

"But you're guessing, right?" Adin asked me. "You don't really know for certain."

He was asking for something that would demonstrate that a largest prime was a logical impossibility. I thought back to the first proof I had done with Bauji showing that any repeated three-digit number must be divisible by 7, 11, and 13. But in this case I had no idea where to start. Somewhat reluctantly, I admitted this to Claire.

"I didn't either," Claire said, "but when I saw the proof I was amazed by its simplicity and . . . grace." She had settled on "grace" after a pause, searching for exactly the right fit for her thought. "I'll explain this the way I first understood it." She put down her drink (ice water with a slice of lemon) and pulled out a pen from her purse. "Let me ask you a few questions before we get down to the proof," she said. "First, is 2 × 5 divisible by 2?"

Of course it is. "2 × 5 is 10 and 10 is divisible by 2." This was obvious; I wondered where Claire was going with this.

"Okay," she said. "Is 2 × 8653 divisible by 2?"

I had to pause, but only for an instant. "Yes, 2 times anything is divisible by 2."

"Why is that?"

"It's true by definition. If you multiply something by 2, the product must have 2 as a divisor."

Claire nodded. "Fair enough. Now, is 7 × 4000 divisible by 7?"

"Yes," I said. "It is, for exactly the same reason."

Claire ignored my "this is obvious" shrug. She was laying a foundation for something.

"Now let me ask you," she said, "is (7 × 4000) + 1 divisible by 7?"

This was the first question I had to think about. 7 × 4000 was 28,000 and the plus 1 made it 28,001. 28,001 divide by 7 would leave . . . 1 as a remainder. Of course it would! 7×4000 was evenly divisible by 7, so $(7 \times 4000) + 1$ would have to have 1 as a remainder when it was divided by 7.

Claire had been watching my face and knew that I had the answer before I said anything. "Good," she smiled. "Now I get to the main idea. We're interested in showing that there is no largest prime. Assume for a second that there is a largest prime."

That seemed an unexpected thing to do. "Why?" I asked.

"You'll see," said Claire. "Now this largest prime, let's give it a name. Call it P."

P seemed like a reasonable name for a largest prime, but I still wasn't sure why you'd assume the existence of something you wanted to prove didn't exist.

Claire waved off my objection and showed me a curious product she had written: $N = 2 \times 3 \times 5 \times 7 \times 11 \times \cdots \times P$.

"I've taken all the prime numbers and multiplied them together. I'll get some number that is much, much larger than P. Call this number N."

So she had assumed there was such a thing as the largest prime. She had named it P. Then she had taken all the prime numbers and multiplied them together and called the product N.

"Is N divisible by 2?" she asked.

Of course it was. $N = 2 \times$ (something), so it was divisible by 2.

"Is it divisible by 3?" she asked, after I had nodded on her first question.

Since N was also equal to $3 \times$ (something), it was also divisible by 3. In a flash I realized that N was divisible by every single prime number. It was constructed that way. "Claire, N is the product of all primes. So it is divisible by all primes."

She nodded. "What about $N + 1$? Is it divisible by 2?"

Her tone told me that we were at the critical point of the argument. I thought about this question carefully. If N was divisible by 2, $N + 1$ would have 1 as a remainder when it was divided by 2. For example, 14 is evenly divisible by 2, but 15 divided by 2 gives 1 as a remainder.

Then the warmth of a sweet Aha moment spread through me. I suddenly realized that $N+1$ divided by any prime would have 1 as a remainder. And then the logical extension was that if no prime divides $N+1$, then $N+1$ would itself have to be a prime, for if a number has any factors, then it must have prime factors.

I couldn't quite close the proof, though. "Wait a minute, Claire! How could $N + 1$ be a prime? You had assumed that P was the largest prime!"

"Ravi, that's the whole point. On the one hand, N + 1 must be a prime because no prime divides it, and on the other hand, P was assumed to be the largest prime. So N + 1, which is much larger than P, cannot possibly be a prime. We have a contradiction. N + 1 is shown to both be a prime and not be a prime at the same time. This cannot be, and there can only be one culprit for this state of affairs."

I could not believe I hadn't seen it at the very start. "The culprit," I told Claire, "is the assumption that there can be a largest prime. As soon as you assume P exists, you get a contradiction. So P cannot exist. There cannot be a largest prime, and therefore the primes must go on forever!"

I held out my arm for a high five. It was not something I was used to doing, but the gesture felt neither forced nor unnatural. Claire was surprised for a second, but then went along with the moment.

Meanwhile, Adin was staring at Claire's notebook. "That's pretty good," he said, mostly to himself.

"*Now* are you absolutely certain that there is no largest prime?" she asked, teasing a little.

"I am. If there were, we'd have a contradiction. There cannot be a largest prime."

Claire smiled, enjoying the victory. "This is an idea that is not relative. It's an idea that people from everywhere would have to believe, an idea of the type whose existence was denied by your Sophists."

I wasn't quite sure what to make of that. That there can be no largest prime seemed like an absolute truth to me, but whether similar methods could be applied to the more philosophical questions of life seemed doubtful.

Peter came in before we could delve further into the question. "Ravi I want to introduce you to this guy from Goldman Sachs. He could probably get you an interview next week. And c'mon you guys, it's time for cake."

"Oy, all this eating," said Claire. "I'll have to go for an extra-long run tomorrow."

• • •

Carol wasn't at the library on Friday morning but she had left a note for me. Her strokes were long and firm, angling to the right:

> RAVI,
> The FedEx package arrived this morning. All the microfiche is there. I've left them at the information desk. Ask someone if you

need help getting started with the microfiche reader. It's on the third floor.

Happy hunting!
C

But the microfiche reader was broken and it could not be repaired until the following Monday. So it was after a restless weekend and a sleepless Sunday night that I set my eyes on the January 1, 1919 edition of *The Morisette Chronicle*. "Happy New Year, Morisette!" said the first headline in an Old English Text font.

In about half an hour I had read through the first week of newspapers. Not unexpectedly, there was no mention of Bauji yet, but I was getting a feel for what the *Chronicle's* editors deemed newsworthy. The four stories on the front page of the January 8 edition seemed to be fairly representative. The lead banner read, "Speeding automobile collides with horse on Elm Street." It was an account of an unfortunate, albeit apparently minor accident. There was a dramatic sketch of a horse with its front legs raised in the air, as if he was ready to hammer down on the oncoming automobile. In a concluding paragraph, the writer clearly sided with the animal over the car:

> God made the horse to provide a way for man to get about. The horse has been doing its job honorably and speedily for hundreds of years. Why a certain number of misguided maniacs would eschew the nobility of a horse for the clutter of an automobile is beyond this writer's understanding.

The second headline read, "Morisette Paper to Lower Production This Spring." The accompanying article discussed the possibilities of further job losses in the mill. The failure of the paper mill to be a reliable employer seemed to irk the reporters of *The Chronicle* to distraction. Almost every day the issue was raised in some fashion. The headline in the rightmost column struck a social note: "Businessman Jerry Adams weds Cynthia Furyk." There was a faded photograph of the couple. They sat tensely, a respectable distance apart from each other, and looked straight ahead at the camera wearing stern expressions. The article noted that Jerry Adams was one of Morisette's finest hunters and his new bride excelled in tailoring. Toward the bottom of the page the headline announced, "Mayor leads fundraising efforts for local church." The article struck on what seemed to be the two dominant themes of the newspaper: local politics and religion.

The one concession *The Chronicle* made to life outside Morisette was a section titled "The World in Brief." It was a quarter of a page, relegated to the back of the newspaper. However, Morisette's definition of "The World" appeared mostly to be limited to New York City. The editors took apparent delight in including stories that painted the city as an unwholesome, even dangerous place. There were stories entitled "Appalling Rise in Prostitution in Brooklyn," "Another Murder in Harlem," and even "Rampant Godlessness"—a story that discussed declining attendance in New York City churches. The reporter gladly pointed out that Morisette had resisted this trend:

> Happily this devil-may-care secularism has stopped short of Morisette's city walls. Nearly all of Morisette's adults attend church regularly and Miss Morison's Bible study classes remain as packed as ever.

New York existed mostly as a counterpoint to Morisette and the world outside New York may as well not have existed at all. Even though the Great War was reshaping Europe, there appeared to be nary a mention of it. But if it happened in Morisette, however trivial, it was news.

In about four hours I had read through all of January's papers. There had been no mention of Bauji. I struck pay dirt later that afternoon. The story I was looking for was on the back page of the March 8 edition:

HINDOO VISITS MORISETTE

MORISETTE — Rarely does the arrival of a single person generate the kind of curiosity aroused by the presence of a Hindoo gentleman in our midst. Morisette is already abuzz with news of Mr. Vijay Sahni, and the citizens of this town, who hitherto have prided themselves on not expressing the slightest interest in the doings of their fellow man, have taken to the most intense scrutiny of his activities.

Mr. Sahni arrived here last week after a stay in London. While he has been referred to variously as a visiting prince or a magician, *The Morisette Chronicle* has learnt that the truth is less exotic. Mr. Sahni is on a visit to this country to pursue mathematical research at Morisette University with our own Dr. Shirer. He has also said he would like to undertake a study of our customs, which no doubt are as strange to him as the life of a Hindoo is to us.

Already Mr. Sahni's presence is being felt in different ways. Yesterday, at the public square, a new chapter was written in the tradition of frank public debate that serves to distinguish our town. Mr. Taylor, a frequent and eloquent Morisette speaker, launched a strong attack on the policies of our English friends in India, from whence Mr. Sahni hails. He compared the fate of this vast subcontinent, with its lack of freedom, to that of the American colonies one hundred and fifty years ago. This provoked unfavorable comment from Mr. Sutton, an audience member, who expressed grave doubts over the comparison between a Christian nation and the heathen population of India.

Mr. Hennings, who is Mr. Sahni's landlord, informs us that the Hindoos are worshippers of the great god of the trinity. The Hindoos believe the trinity is responsible for the creation, preservation, and destruction of this world, which has been created and destroyed several times in the past.

Some Hindoos, Mr. Henning tells us, begin their day by paying homage to the sun, which for them embodies the trinity on earth. The sun, they claim, is recreated at dawn and preserves the earth through the day, only to be cast to the shadows in the evening. Each day is for them a model of the eventual fate of our world.

Some people have reported observing Mr. Sahni looking at the sun while walking outside. The religious significance of this remains unknown.

Looking at the sun was one of Bauji's old habits. He used to be somewhat prone to migraines and said that sunlight kept away the headaches. But that was a minor aside; the big news was that Bauji had indeed come to Morisette, apparently, in the words of *The Chronicle,* to "pursue mathematical research." My guess was that this Dr. Shirer was a local math professor who had read some of Bauji's papers and had invited him to visit Morisette University.

The confirmation of his arrival increased my appetite for more information. I didn't have to look too long. Only two days later there was another reference to Bauji:

A SURPRISE VISIT

MORISETTE — The appearance of a brief report on the presence of a Hindoo gentleman in our midst has yielded this newspaper the unexpected pleasure of hosting Mr. Vijay Sahni at our office. Mr. Sahni dropped in after reading the item that appeared on March 8, 1919 to "introduce himself and dispel misconceptions that may have arisen in the course of my stay in Morisette."

In person, Mr. Sahni comes across as a well-educated gentleman of means. In appearance, he does not differ much from the Italian or Greek immigrants who have been landing on our shores in large numbers, but his knowledge of English would put a native of this country to shame.

This comes as no surprise after he informs us that he had been a student of mathematics at Cambridge before he arrived in Morisette. He says that aside from furthering his mathematical studies, he is here to experience firsthand the atmosphere of independence and free thought that America is renowned for, and names Jefferson and Lincoln among the men he admires most.

He is far more reticent on the subject of his own nation. He chose not to dwell on questions about snake charmers that some eager visitors to the office directed at him and laughed away the suggestion of any connection with royalty. He informs us that he hails from the northern province of Punjab, which capitulated to the Britons only fifty years ago, and has lived and studied in the capital, Lahore.

He was full of praise for the American way of life and the right every free man enjoys to live his life as he chooses. Speaking in particular of Morisette, he expressed the greatest appreciation for our own local tradition of public debate so forcefully exercised at the town square.

And while we extend a warm hand of friendship to this Hindoo gentleman, we will also do well to remember that it is the influence of a Christian nation and the unstinting effort of the Caucasian man that have brought him the education that stands him in such good stead in our midst.

I wondered how Bauji would have reacted to the last paragraph. He was a proud man and, unlike many educated Indians of his time, he did not feel that he owed his success to the good graces of the White man. I wondered whether this was the issue that had somehow led to his imprisonment.

Two days later, the puzzle was solved. The answer screamed in two-inch high headlines on the front page of the April 3 edition of *The Morisette Chronicle*:

OFFENSIVE SPEECH BY HINDOO

Morisette — This quiet hamlet of Morisette has suddenly been beset with unwanted turbulence following an offensive speech by the Hindoo gentleman, Vijay Sahni, who has figured in these pages on two previous occasions.

According to a bystander who prefers to remain unnamed, the incident took place around five o'clock in the evening, when several folks had gathered to hear another variation of Mr. Patterson's impassioned defense of the Christian way of life. Spurred, perhaps, by the presence of a Hindoo in our midst, Mr. Patterson chose to point out the superiority of the Christian way of life over the superstitious outlook of the Oriental, in particular the Hindoo.

Mr. Patterson, as he is wont to do, had warmed up to his subject, describing the superstitions that prevail upon and corrupt the teeming population of the Hindoo nation, when Mr. Sahni happened to pass by the square. Bystanders claim that Mr. Patterson did not realize Mr. Sahni was present.

After Mr. Patterson had concluded, Mr. Sahni came forward to address the audience. According to Mr. Patterson, Mr. Sahni then proceeded to launch an insulting attack against Christian belief and the authority of the Bible. In particular, Mr. Patterson stressed that in defending the Hindoo way of life, not only did Mr. Sahni question the Biblical version of the creation of the world and propound the Darwinian heresy, he also cast serious doubt on the text of the Bible and upon Protestant teachings. Among his many offensive statements, Mr. Sahni characterized Christianity as a "procession of fools who have given up on reason." He is said to have remarked that America, the land of "rationality and objectivity," had no room for "the darkness of illogic."

According to another member of the audience, "Mr. Sahni strode onto the podium in some anger but he spoke with what appeared to be a calm logic. Beginning with an offensive aside on the virgin birth and the story of creation in the Bible, he related the origins of Protestant thought, dwelling on the original inspiration and commented that current Christian practice is divorced from anything resembling 'defensible spirituality'. In the course of this speech," Mr. Patterson adds, "he made several uncharitable references to the Protestant form of worship, service, and sacrament."

At this point Mr. Sahni was severely heckled by the audience, having clearly exceeded the bounds of free speech as understood in our town. After Mr. Sahni left, several of the audience expressed considerable resentment at his speech.

The next day's edition contained news of Bauji's arrest.

HINDOO ARRESTED

MORISETTE — Mr. Vijay Sahni was arrested today by Morisette sheriff Craig Johnson on charges of blasphemy. The rarely invoked law has been applied after complaints were lodged with the sheriff over the uncharitable references to Christianity made in the course of a speech by Mr. Sahni at the town square yesterday.

The sheriff told *The Morisette Chronicle* that, today, as word of Mr. Sahni's anti-Christian speech spread, it became imperative that he take action against Mr. Sahni in the interest of maintaining peace. Upon due consideration by the local prosecutor, Mr. Daniels, it was decided that Mr. Sahni be booked under the state's Blasphemy Law (see box). However, a deposition was required to enable the arrest. Mr. Patterson, whose speech caused the Indian's outburst, was present at the Sheriff's office to make the deposition. The Sheriff's report of Mr. Patterson's deposition is produced below:

> Mr. Patterson, on his oath, said that on the evening of April 2, 1917, he was at a public meeting held in the gardens of Mayberry Park in the town square section on the westerly side of Maple Street, Morisette, N.J. The meeting was addressed by one Vijay Sahni; and said Sahni publicly blasphemed the holy name of God by denying and contumaciously reproaching the being and existence of God and of the scriptures as contained in the books of the Old and New Testament by saying birds and fish were made out of lower organisms and said other nonsense trying to prove the scripture false and making fun of and ridiculing the Bible. He also made diverse other remarks, offensive to the fabric of Christianity. Therefore the deponent prays that said Sahni may be arrested and dealt with according to the law.

Mr. Sahni was booked under the Blasphemy Law exactly twenty-four hours after the events that took place at the town square. He is said to have expressed considerable surprise at his arrest. However, he offered no resistance.

Supporting the article was a statement of New Jersey's Blasphemy Law. I'm told that the law stands unchanged to this day.

SPECIAL FEATURE: NEW JERSEY'S BLASPHEMY LAW

MORISETTE — If any person shall willfully blaspheme the holy name of God, by denying, cursing, or contumeliously reproaching his being or providence, or by cursing or contumeliously reproaching Jesus Christ or the Holy Ghost, or the Christian religion or the holy word of God (that is, the canonical scriptures contained in the books of the Old and New Testaments), or by profane scoffing at or exposing them, or any of them, to contempt and ridicule, then every person so offending shall, on conviction thereof, be punished by a fine, not exceeding two hundred dollars, or imprisonment at hard labor not exceeding twelve months, or both.

The following day's newspaper contained two editorials on the front page. The one on the left had been reproduced from the previous day's *New York Times*. The one on the right appeared to be a rebuttal from the editors of the *Morisette Chronicle*.

EDITORIAL: REPRODUCED FROM *THE NEW YORK TIMES* WHAT ABOUT FREE SPEECH?

MORISETTE — The news from Morisette is not good. It is difficult to believe that in this century, in this country, just thirty miles from New York, a man can stand trial for blasphemy. It does not help that the accused is a Hindoo who has come to these shores to escape the yoke of British rule, a yoke that had also weighed down our ancestors.

What is even more shocking is the way in which the principle of free speech, one of the cornerstones of our nation, has been sacrificed on the altar of religion. No one makes the case that faith should be subjected to scorn, but inherent in the right to free speech is a tolerance that is designed to err on the side of excess. The danger lies precisely in attempting to limit this right; one man's reasonableness may be another man's extremism.

The local sheriff, in the interests of maintaining the peace, has arrested the accused for blaspheming the Bible and the Christian faith. This cannot be ignored as an isolated event, in a small town in America. The hurt caused to the sentiments of many in our country should not blind us to the dangers of proceeding with this trial. For centuries, our nation has been a safe haven for those who have fled Europe in face of persecution for their beliefs. And, however heterodox those beliefs, they have continued to thrive in America.

The unrestricted right to free speech, so central to our political life, has been upheld time and again by the courts of our nation. A trial on the grounds of blasphemy can only seek to circumscribe this right. For this very reason, the idea of an arrest under this archaic law of blasphemy and a trial on this ground is untenable. Ever since the issue has surfaced, the governor of New Jersey has failed to make his position clear. In the meantime the damage done to the democratic traditions in his state cannot be ignored. It is time that the governor intervene to prevent the mockery of a trial that will call into question one of the most basic precepts of our polity. The governor must act.

THE MORISETTE CHRONICLE'S REBUTTAL TO THE TIMES
A SENSE OF OUTRAGE

MORISETTE — There is nobody in Morisette who is not familiar with the events of the past two days. The hospitality extended by this town has been grossly abused by a stranger and the sheriff has chosen to act in the best traditions of the law. What surprises us, and the citizens of Morisette, is the manner in which some in the outside world have reacted to this event.

Morisette is a small town, built by the honest toil of hardworking citizens. Towns such as ours have made America great and it is here that the American way of life was born and bred. The spirit that has made these shores the beacon of the free world was nurtured in hundreds of towns that resemble Morisette in spirit.

The Christian tradition is central to this way of life. To call it into question is to question the idea of America itself. And what nobler idea has ever been conceived—an idea that has let millions savor the joys of freedom, free from the persecution that was prevalent in the Old World?

It is for this very reason that the law must be vigilant. While our Founding Fathers envisaged great latitude in the tradition of free thought and speech, certain limits must exist which respect the spirit of our nation. While Mr. Sahni is welcome to his beliefs, he is not welcome to enjoy our hospitality and, in the confines of a public space nurtured by us, abuse the hospitality that Morisette has so generously extended him.

The words he has used in reference to Christianity have hurt every citizen of this town. Our way of life is precious to us and it is inextricably linked to Christian thought. And today, as newspaper after newspaper, from the cities of New York and Boston report this arrest—sometimes in highly negative terms—they fail to understand the very basis of a way of life that makes their existence possible. Without Morisette, there is no New York.

We ask you, if it so hurts Mr. Sahni to live on these shores, what would have happened if he had tried to speak in similar terms in the Islamic nations that border his country or in the towns of Europe. It is only here that he will live to realize the errors of his ways and learn to appreciate the spirit of the New World.

So there it was, the full catastrophe. I was getting late for Nico's class and it was time to leave. As I walked out of the darkness of the microfiche room into the main quad amidst the cyclists and the skateboarders, Bauji's story seemed remote and unreal to me. Yet it had happened, in this country, and not all that long ago.

• • •

Nico started off the class with an apparently simple-looking infinite sum: $S = 1 + 1 + 1 + 1 + 1 + 1 \ldots.$

"What is the value of S?" he asked.

It was tempting to say "infinity" (and many in the class did), since the string of 1s ran on forever. But last week, Nico had shown that infinity is not a number in the sense that, say, 37 is a number. So to me, it seemed incorrect to say that $S = \infty$.

Finally, Adin said that S had no value. "It's a meaningless question."

Nico laughed at that. "You're right, Adin. It is meaningless to ask for a value for S because the infinite sum we're considering diverges. One of the ways a sum may diverge is if, by adding enough terms, its value can be made as large as we like. So if I challenged you to show me that S was larger than 10 billion, you'd just add up the first 10 billion and 1 terms of the sum. No matter what threshold I gave you, you could make the sum exceed it; that's one of the hallmarks of a divergent sum. Other divergent sums, instead of increasing without bound, may unendingly oscillate without telescoping into a particular value. The series of alternating 1s and –1s we examined in the last class is an example of this phenomenon."

Nico briefly drummed the side of the table with his fingers. A simple 1-3-1 beat, to collect his thoughts. "But not all infinite sums diverge. Last week we saw a sum related to Zeno's paradox that I told you converges to 4." He went to the blackboard and wrote: "I claim: $2 + 1 + 1/2 + 1/4 + 1/8 + 1/16 + \cdots = 4$."

"But a mere claim is not enough," Nico continued. "In mathematics we can claim nothing without proof. Only proof provides truth."

I couldn't see Adin's face from the back, but noticed that he was rocking forwards and backwards ever so slightly, the movement beginning from the base of his spine. I had seen him do this at the crucial points in last week's lecture as well. He told me later that the had acquired the habit in his childhood synagogue, watching Jewish scholars davening. "It helps me concentrate," he told me later.

Nico proceeded to remind the class about "Claire's beautiful but incorrect argument, where she lined up the terms and subtracted most of them out." He wrote her equations on the blackboard.

$$
\begin{array}{rl}
\text{Sum} = & 2 + 1 + 1/2 + 1/4 + 1/8 + 1/16 + 1/32 + \ldots \\
-\ 1/2 * \text{Sum} = & \phantom{2 +{}} 1 + 1/2 + 1/4 + 1/8 + 1/16 + 1/32 + \ldots \\
\hline
= 1/2 * \text{Sum} = & 2 + 0 + 0 + 0 + 0 + 0 + 0 + \ldots.
\end{array}
$$

$$\Rightarrow 1/2 * \text{Sum} = 2$$

$$\Rightarrow \text{Sum} = 4$$

"The problem with this argument," said Nico, looking at Claire, "is that nothing gives Claire the right to multiply both sides of an infinite equation by 1/2, and furthermore, nothing gives her the right to subtract out an infinite number of terms like that. Even though the method happens to provide the correct answer, finite methods may not be applied to infinite quantities."

It seemed to me that Nico's objection was almost legalistic. There were some laws of multiplication and subtraction, and those laws could safely be applied to equations with a finite number of terms. As far as I knew, we had no laws to deal with infinite sums.

"The way around this difficulty," said Nico, "is for us to limit our attention to a finite number of terms." He went back to the blackboard:

$$\text{Sum} = 2 + 1 + 1/2 + 1/4 + 1/8 + 1/16 + \cdots.$$

"What is the seventh term after 1/16?"

"1/32," said Peter.

"Right. What is the tenth term?"

1/32 was the seventh term, so 1/64 was the eight, 1/128 was the ninth, and 1/256 was the tenth. The whole class went through the same calculation and Peter was the quickest on the draw.

"Right again," said Nico. "What is the 100th term?"

This was harder. It would be silly to even try to count off the terms one by one. Despite my preoccupation with *The Morisette Chronicle* I felt myself pulled into the problem and looked for a way to attack it. As soon as I organized my doodles into a table, I saw the pattern:

Term #	*Value*	*Value Expressed as Power of 2*
3	1/2	$1/2^1$
4	1/4	$1/2^2$
5	1/8	$1/2^3$
6	1/16	$1/2^4$

It certainly seemed that the 100th term would be $1/2^{98}$. I was about to volunteer the answer but Claire was faster.

"Correct. Well done, Claire," said Nico. He then pushed us to generalize the pattern. "What is the nth term, where n is any integer?"

This, too, seemed clear enough. "It would be $1/2^{n-2}$," I replied.

"Excellent," said Nico. "Now let us write out the sum of the first n terms," he said going to the blackboard:

$$Sum_n = 2 + 1 + 1/2 + 1/4 + 1/8 + 1/16 + \cdots + 1/2^{n-2}.$$

"Notice that instead of having an infinite sum we now have a finite sum, although since we have complete control in assigning a value to n, we could include as many terms as we like. For example, here is the sum of the first 1000 terms," he said, going to the blackboard:

$$Sum_{1000} = 2 + 1 + 1/2 + 1/4 + 1/8 + 1/16 + \cdots + 1/2^{998}.$$

"All I've done here is to replace n by 1000. Equivalently, you could have n equal to a billion or a trillion. What's more important is that by considering the sum of the first n terms we have a finite sum instead of an infinite sum. So we're perfectly justified in applying Claire's elegant method of multiplying both sides of the equation by 1/2 and then subtracting terms. Let's see how things work out." With the exception of having concluding terms at the end, Nico repeated Claire's demonstration from last week:

$$Sum_n = 2 + 1 + 1/2 + 1/4 + 1/8 + 1/16 + 1/32 + \ldots + 1/2^{n-2}$$

$$- 1/2 * Sum_n = 1 + 1/2 + 1/4 + 1/8 + 1/16 + 1/32 + \ldots + 1/2^{n-2} + 1/2^{n-1}$$

$$= 1/2 * Sum_n = 2 + 0 + 0 + 0 + 0 + 0 + 0 + \ldots + 0 - 1/2^{n-2}$$

$$\Rightarrow 1/2 * Sum_n = 2 - 1/2^{n-1}$$

$$\Rightarrow Sum_n = 4 - 1/2^{n-2}$$

"The sum of the first n terms is $4 - 1/2^{n-2}$. To make this concrete let's look at an example: The sum of the first 100 terms is $4 - 1/2^{98}$, which is just a tiny smidgen less than 4."

I did a quick calculation to see what sort of smidgen Nico was talking about. $1/2^{10}$ is around 0.0009, $1/2^{98}$ is *much, much* smaller.

"The larger you make n, the smaller you can make $1/2^{n-2}$," said Nico. "And that is the key idea behind the concept of limit. In fact, we say that the limit of $1/2^{n-2}$ tends to 0 as n tends to ∞. What we're saying is that one can make $1/2^{n-2}$ as close to 0 as one wants just by making n large enough. No matter how small a quantity you name, I can find n such that $1/2^{n-2}$ is closer to zero than the quantity you have named. Thus, as the number of terms increases without bound, the sum converges to 4. It is in this sense that $4 = 2 + 1 + 1/2 + 1/4 + 1/8 + 1/16 + \cdots$."

Nico was pacing now and his voice was louder. "Simple as it seems, ladies and gentlemen, we have done some deep and powerful mathematics here. We have taken an infinite series and, using a logically justifiable approach, we have proven that the sum must equal 4. We have not guessed; we have not waved our arm over the details. Instead we've taken a systematic approach, each step of which is grounded in reason, and we've come up with our answer. Along the way, we were even compelled to formulate the notion of limit. But the upside of all this trouble is that we can now claim certainty. We can be confident that our infinite sum converges to 4. Even though we are dealing with an infinite sum, a difficult customer at first glance, we have tamed it and solved its mystery." He looked at the equations on the blackboard and nodded to himself. "And this completes our answer to a question that may have troubled Zeno many centuries ago."

I could see Adin shake his head. He was convinced that Zeno was too smart to have been stumped by infinite series. Perhaps, as Nico had suggested last week, he had in fact been troubled by a different issue entirely.

Nico noticed Adin's disagreement. "I said it *may* have troubled Zeno," he said smiling. Then turning to the room, he said, "I brought up the infinite series in the context of Zeno's paradox to motivate us to think about infinite series in the first place. The convergence of the series—and not Zeno's paradox—has been our main point. And know that this is really cool stuff we've done today. Should some of you happen to take calculus, you'll again run into the notions of limit and convergence. But this is as far as we'll go for now."

There was a collective release of tension in the room. Nico made us feel we had participated in developing the answer, not just watched it unfold. Looking back I realize that he did this by asking small, simple questions (that we could chew on and digest), and then using them as building blocks to resolve bigger, more complex puzzles. It's a great way to teach anything.

"We're out of time for today but I hope you will play with these infinite series at your leisure," said Nico. "The underlying mathematics here is one of the most beautiful creations of the human mind. It is no less elegant than a Mozart concerto, or a Vermeer painting. I particularly encourage you to read about the harmonic series. Try to figure out if it converges or diverges without looking at the answer."

"What's the harmonic series?" PK asked.

Without a word, Nico wrote the series on the board:

H = 1 + 1/2 + 1/3 + 1/4 + 1/5 + 1/6 . . . .

Perhaps to forget about Bauji in Morisette, I immediately lost myself in finding the answer. When I looked up, nearly two hours had gone by and the classroom had long emptied. But I was happy, for I had cracked the case, and the solution was so simple, yet so elegant. Bauji would have been thrilled to see it.

I did swing by Nico's office to tell him, but he was not there. So I sketched out my solution and slid it under his door.

• • •

NICOLE ORESME JOURNAL ENTRY, 1354 AD

It has been a calamitous few years. First our armies were defeated at Crecy, then the Black Death took the lives of untold thousands, and now I hear dark rumours of the Turks taking Gallipoli and gaining a foothold in Europe. Through these turbulent times Mathematics has been a great solace for me. I've enjoyed thinking about some of the puzzles that have confounded some of the most learned men in Paris. In all the ugliness of life, Mathematics can sometimes provide brief glimpses of great beauty.

Just yesterday, for example, I created the solution to a delightful little problem suggested to me by a bright young student at the College of Navarre. The student, Sebastien, showed me an infinite sum that he claimed was in some sense divine. Here is his series:

1 + 1/2 + 1/3 + 1/4 + 1/5 . . . .

As you can see, each term of the series is less than its predecessor, yet Sebastien claimed that his series did not converge to a finite value!

This is an extraordinary claim. It has been known since the time of the ancient Greeks that series whose terms decrease to zero do, in fact, have a finite sum. For example, Aristotle knew that 1 + 1/2 + 1/4 + 1/8 + 1/16 . . . = 2. There are many other such examples. Decreasing terms in an infinite series has been a sure signal of convergence.

Yet Sebastien was adamant that his series did not converge. When I asked him for a demonstration he readily admitted that he had none. He was basing his claim on "mathematician's intuition." If anyone else had provided this sort of rationale, I would have been immediately dismissive, but Sebastien does indeed have a remarkable ability to spot mathematical truth. He is seldom wrong about such things.

On coming home, the first thing I did was to check the monastery library for references to the series. Yet I found almost nothing. The one place I saw the series mentioned was in an old Italian text that claimed the series converged since the terms "decreased to almost nothing very quickly."

My first few calculations also tended to support the convergence hypothesis. The sum of the first 50 terms of the series (relatively) quickly climbs to about 4.5. The next 50 terms add up to only about 0.7 and the 50 after add up to a hair more than 0.5. This seems to be the behaviour of a convergent series. (Fortunately I had a group of students doing the rather tedious calculations!)

But late last night I was inspired by an idea that simplified everything. I was able to show that despite its slow growth the series does, indeed, diverge—Sebastien was right after all!

The main idea was that if I could show that Sebastien's series was greater than or equal to another series that we know diverges, then it itself must diverge. In particular, we know that the series 1/2 + 1/2 + 1/2 + 1/2 + 1/2 . . . diverges since it exceeds any possible finite value. I was able to show that Sebastien's series is greater than this sum of an infinite number of halves. Mathematically:

$$1 + 1/2 + 1/3 + 1/4 + 1/5 + 1/6 + 1/7 + 1/8 \ldots \geqslant 1/2 + 1/2 + 1/2 \ldots.$$

This does not seem likely at first sight. After all most of the terms of the series on the left are actually less than half. But watch this:

$$\begin{aligned} &1 + 1/2 + 1/3 + 1/4 + 1/5 + 1/6 + 1/7 + 1/8 + 1/9 + \cdots \\ &\quad = 1 + 1/2 + (1/3 + 1/4) + (1/5 + 1/6 + 1/7 + 1/8) \\ &\qquad + (1/9 + 1/10 + 1/11 + 1/12 + 1/13 + 1/14 + 1/15 + 1/16) + \cdots. \end{aligned}$$

All I've done is to group the terms of the series. But do you notice something?

$$\begin{aligned} &(1/3 + 1/4) > (1/4 + 1/4) = 2/4 = 1/2, \\ &(1/5 + 1/6 + 1/7 + 1/8) > (1/8 + 1/8 + 1/8 + 1/8) = 4/8 = 1/2, \\ &(1/9 + 1/10 + 1/11 + 1/12 + 1/13 + 1/14 + 1/15 + 1/16) \\ &\quad > (1/16 + 1/16 + 1/16 + 1/16 + 1/16 + 1/16 + 1/16 + 1/16) \\ &\quad = 8/16 = 1/2. \end{aligned}$$

Each grouping in the parenthesis is greater than 1/2. The next parenthesis will have 16 terms, the one after that will have 32. The beauty of an infinite series is that we never run out of terms! You can always get enough to exceed 1/2. Which is why I can say

$$1 + 1/2 + 1/3 + 1/4 + 1/5 + 1/6 + 1/7 + 1/8 \ldots \geqslant 1/2 + 1/2 + 1/2 \ldots .$$

Since Sebastien's series is greater than a divergent series, it itself must be divergent! And therefore Sebastien was right. The series does, indeed, diverge even though its terms successively decrease. What an unusual animal!

I have a strong desire to immediately inform Sebastien of my discovery. But unfortunately there is no device that can magically transmit my words to the other end of Paris. I will have to wait until next week, when I will see him at the college.

• • •

That evening I got an e-mail from Nico which I have saved to this day.

Ravi,
Sorry I missed you when you came by to drop your solution to the Harmonic Series question. Your analysis is absolutely spot-on. It may interest you to know that an argument very similar to the one you have developed was first completed around 1355 by one Nicole Oresme, a French scholar. As you have done, he showed that you can break up the series into clusters, each of which is greater than 1/2.

There are many fascinating things related to the harmonic series. Since you are so obviously interested in this subject, I will sketch some of them out: First consider the series of the reciprocals of the prime numbers. Since (as you may know) there are an infinite number of prime numbers we can be sure that this is in fact an infinite series (i.e., it goes on forever, and does not have a last term):

$$1 + 1/2 + 1/3 + 1/5 + 1/7 + 1/11 + 1/13 + \cdots .$$

Notice that in some sense this series is a small part of the harmonic series. Since so many of the terms of the harmonic series are missing (e.g., 1/4, 1/6, 1/8, 1/9) you may think that the series converges. But it does not! It can be demonstrated (using a semicomplicated argument) that this series diverges. I remember being completely surprised when I first saw this result.

My second fascinating factoid has to do with twin primes. These are consecutive odd numbers, both of which are primes. 3 and 5 are twin primes, as are 11 and 13 (7 and 9 are not twin primes since 9 is not a prime). Almost everyone believes that there is an infinite number of twin primes. But despite the best efforts of the most brilliant minds throughout history, this conjecture has resisted proof. It has been shown, however, that the series of the reciprocals of the twin primes does converge.

So 1 + 1/3 + 1/5 + 1/11 + 1/13 + 1/17 + 1/19 . . . is a convergent series. Now, it is possible that this is, in fact, a finite series (if the twin prime conjecture is wrong), in which case its convergence is not at all a surprise since all finite series converge to a finite sum. But if, as everyone expects, this is an infinite series, it is truly intriguing that it converges. And, in case you're wondering what value the series converges to, it has been proven that the sum of the reciprocals of the twin primes is about 1.9021605 . . . (this is called Brun's constant).

In some sense, then, the primes are a "large" subset of the integers (since their reciprocal series diverges) and the twin primes are a "small" subset (since their reciprocal series converges). We will come up with a more useful way to compare sets a little bit later in class.

Lastly, Ravi, I want to leave you with another convergent series: the reciprocal of squares. You'll never believe what it converges to:

$$1 + 1/4 + 1/9 + 1/16 + 1/25 + 1/36 + 1/49 + \cdots = \pi^2/6.$$

Yes, $\pi^2/6$!! That is the same π that you are familiar with: the ratio of a circle's circumference and diameter. How did it get into this equation? What could a ratio pertaining to circles have to do with the sum of reciprocals of squares? Is this not truly miraculous?

Sometimes when I look at this equation with fresh eyes, I'm amazed and in awe all over again. Equations like this represent why I fell in love with mathematics. There are so many unexpected connections, so much order when you would expect none, a mostly hidden tapestry into which we get a few limited glimpses through the efforts of our brightest minds.

Who made these connections? Why do they exist? I can only say that God must be a mathematician.
Nico

CHAPTER

3

> An absence of the Lord is the presence of Satan. A heathen has blasphemed the Lord and he shall rot in Hell. But should we allow him to get away with less on earth? No punishment will be severe enough for the Hindoo Satan in our midst.

So concluded Mrs. Ethan Gardener of 37 Parkinson Lane, in a letter to the editor of *The Morisette Chronicle*.

In the days after the arrest, *The Morisette Chronicle* was transformed from a mild New Jersey town newspaper into a tabloid baying for Bauji's blood. Article after article and letter after letter expressed outrage at what had occurred in the town square. There were reports that the sheriff, in an unprecedented step, had to place armed guards on duty outside the town jail.

Sunday's newspaper went a step further and featured interviews with New Jersey's religious leaders, all of whom thought the foreigner's views were a virulent challenge to Christianity itself. "Christ died on the cross so that mankind might find the path to salvation. It is possible to forgive those who fail to heed His message out of ignorance, but those who blaspheme knowing well the consequences of their acts are no better than those who nailed Christ to the cross."

The few voices that spoke up for the contrarian view were isolated, and buried on the inside pages. And even these invited criticism in the form of further letters filled with bile and invective. One writer effusively described how his own congregation turned against its pastor, who dared to argue,

"Christianity lives and breathes love, and love must even be directed at lost souls such as Mr. Sahni. He is to be pitied, not punished. After all, he is steeling himself against the glory of God, and he cannot feel the bliss of His infinite love."

Dr. Shirer, Bauji's host at the college, wrote in a letter to the newspaper that "however abhorrent the Indian's words are to our sensibility, they are, after all, just words, and in America you can't hang a man for his words." A few days later, in an article prominently placed on the front page, questions were raised about the invitation to Bauji in the first place: "Academic freedom must exist within the limits of virtue. In the name of an amoral striving for science, we cannot and should not let just anyone land on these shores. Among all the mathematicians of this world, how did the University choose Mr. Vijay Sahni? The tax dollars of the American Citizen are not to be wasted on the whims and fancies of a few academics who contribute to the willful affront to our civil society."

Amid the outrage, a strand of indignation at the attitude of New York papers was apparent. A *New York Times* editor must have used the words "provincial" and "backward" in reference to Morisette and the retort betrayed the hurt: "I work hard, take care of my family, go to church on Sunday and volunteer for the fire department. If that is what it means to be provincial, then I am proud to be so. And if clean streets and orderly citizens signify that we are backward, then New York is welcome to its cult of mammon and dens of vice which speak of the progress of Manhattan or Queens.

And so it went on for days. Arguments raged, outrage spread, and the citizens of Morisette demanded action. A whole community was in turmoil because of my grandfather. Yet it seemed odd to me that it was he who would be in the middle of a controversy about faith. The man I knew was somewhat indifferent to religion. He certainly was not devoutly religious, but I did not remember him being vehemently against religion either. He never questioned or argued with people about their faith, and to the best of my recollection, he was silent on the matter of God.

The next piece of real news was reported in the April 17 edition:

JUDGE TAYLOR MAY BE THE GOVERNOR'S WAY OUT

MORISETTE — The governor's intervention seems imminent in the Vijay Sahni Blasphemy case. Even as the controversy refuses to die down it is believed that Governor Williams may be considering options other than a trial or a dismissal. With local sentiment in Morisette ranged against the extreme liberal views being propagated in the New York newspapers, this may help stave off criticism from either quarter.

According to sources close to the governor, it seems more and more likely that he may consider appointing a judge to examine the merits of the case. While several names have been under consideration, the choice is said to be narrowing down to Judge John Taylor (see box).

The option of letting the matter come up for trial is something the governor is anxious to avoid. It is learned that he is wary of any publicity that might show the state in a poor light. Governor Williams has been advised by legal experts that the rarity of such a trial will lead some to making it a test case for the blasphemy law.

With a presidential bid in the offing, the governor is concerned with the criticism being mouthed in the liberal big city newspapers. Neither does he feel that dropping the matter is a realistic option, his advisors admit, in view of the strong opposition that any such step is likely to arouse in the state. A choice either way could end up alienating large segments of this nation.

The appointment of a judge to examine the case is thus a bid to sidestep the dilemma staring the governor in the face. However, it is still not clear how the governor will go about this step. His advisors clearly hope that by granting a large measure of discretion to a judge who is respected for his integrity, the onus of any final decision will not fall on Governor Williams. In fact, it is more than likely that the governor would prefer to go by whatever recommendation is ultimately made by the judge.

Accompanying the article was the following "box" on Judge Taylor:

Judge John Taylor may well be the governor's choice for handling the Morisette Blasphemy case. His advisors have hinted as much and it is clear that such a move will take the heat off the governor. While some citizens of Morisette may feel particularly aggrieved at being denied a trial, it will be a rare man who will oppose Judge Taylor's appointment.

Born and brought up in Morisette, Judge Taylor, in the eyes of most citizens of the town, is an embodiment of how they would ideally like to see themselves. His family was among the earliest settlers in this town, and his father was the pastor of the Episcopalian church on Third Street. Among his forebears are his namesake, who served in the War of Independence and was an associate of Thomas Jefferson.

The judge himself was an outstanding student and has been actively involved in community service in the town. His career has only added to his reputation for probity and fairness. His faith in the Bible is amply reflected in his judgments that are peppered with quotes from the Gospel.

While he has strong Conservative leanings, he has never shied away from taking a contrary stand if he felt the truth was otherwise. These qualities have led many to believe that he may well go on to serve on the Supreme Court.

News of the formal appointment of the judge was published in the following day's edition. Judge John Taylor was indeed the governor's choice.

In the picture on the front page, he appeared to be in his late forties. The photograph seemed to have been taken in his study, for there was an imposing bookshelf behind him. He held a book in his right hand and a cigar in his mouth. His hair was startlingly white and he rested his reading glasses halfway down the bridge of his nose. Despite the graininess of the photograph I could make out from his expression that he didn't want his picture taken, but was willing to tolerate the newspapermen as long as they did their business and left quickly. His thumb was hooked into the book he was reading, and it seemed that as soon as the photographer left, he would spring it open.

In the accompanying article, the Judge refused to comment on the particulars of what *The Chronicle* was now calling the Blasphemy Case. His sole quote, however, did give me cause to hope that Bauji would be dealing with a thoughtful man: "The issues in this case are tangled and complicated. Our laws of free speech are at odds with our blasphemy law; the sensitivities of a community are at odds with an individual's value system. I pray that God gives me the wisdom to see my way through this mess."

The article went on to say that the judge would be making a recommendation to the governor on whether the blasphemy case should be taken to trial. His method of investigation would be to have a series of conversations with the prisoner to determine whether "the Indian's intent was blasphemous, or merely ignorant."

A few days later *The Chronicle* reported that Judge Taylor had had his first conversation with Bauji. However, there were no accounts of the conversation itself. A frustrated reporter noted that while several bystanders had seen Judge Taylor enter and leave the county jailhouse, he had flatly refused to talk about the nature of the conversation he was having with the prisoner. "That is between me and him," he was quoted as saying. Asked if he had drawn any conclusions in the case, Judge Taylor said, "It is too early for any conclusions. I will be meeting with Mr. Sahni again and will render my decision when the investigation is complete. I have no further comments until then." When asked what he and the prisoner talked about he replied, "The conversations are being recorded word for word by Mr. Hanks, the court stenographer. These transcripts will be released to you once this matter is concluded."

The next day's edition commented that the judge appeared quiet and drawn after spending nearly two hours with the prisoner. The editor speculated, "The Christian spirit of Judge Taylor must have been truly tested by the Hindoo's heathen talk." But there were no reports on the actual content of the conversations between my grandfather and the judge. Even Mr. Hanks, the court stenographer was interviewed, but he too maintained a steadfast "No comment."

Once again, there was a lull in real news about the case. Judge Taylor was keeping his findings and the transcripts recording the conversation to himself and in an interview encouraged the citizens of Morisette to "get on with their daily lives." For a few days *The Chronicle* contented itself and its readers by reporting on seemingly insignificant details on what the judge was wearing on his way to the county jail (his grey suit), on how long the conversation lasted (usually 2 to 3 hours) and what his expression was after the conversations ("superficially inscrutable, but internally disturbed," was the beat-writer's oft-reported opinion).

As that Morisette April wore on, the frequency of these reports decreased. The storm and drama, at least for now, appeared to be over. Once again, the citizens of Morisette turned their attention to normal life—the paper mill, the Lincoln High Grizzlies, and the decadence of the New York City public. The blasphemous foreigner was safely locked up.

June 30 was the last edition that Carol had obtained for me. It was after 3:00 a.m. when I finished going through it and I still had no idea what had happened to my grandfather. I only knew that America—my adopted home, the land that I loved—had imprisoned Bauji. On the way out of the library I nodded to the security guard. He had his radio playing and I could make out Ray Charles pining for Georgia.

The next morning I wrote an e-mail to Carol Stern telling her the whole story. I told her about the arrest, the appointment of the judge, and the conversations whose content I did not know. "Can I get transcripts of these conversations and can I get the editions from July 1 to Dec. 31, 1919?"

• • •

"So, Ravi, why do you want to work for Goldman Sachs?"

John Schilling had been interviewing me at the Campus Recruiting Center for the better part of an hour. He had thrown me a series of probing questions on merger strategy, leveraged instruments and valuation theory. The questions were designed not only to test my understanding of the content, but also my ability to think on my feet. Surprising even myself, I found myself responding with reasonable (and, I think, correct) answers to each of his zingers. This last question about why I wanted to work at Goldman Sachs was known in interview circles as a "fit" question. Its purpose was not so much to evaluate whether the candidate was sincere about the content of his answer, but to see if he could generate the "right" kind of verbiage. I knew I was expected to talk about things like Goldman's "market leading presence" and its "people-friendly culture." The point was mostly to demonstrate that I could talk the talk.

Probably because I'd rehearsed the answer and already knew what I would say, I relaxed just a little. Just for an instant I allowed myself to reflect on the actual reasons that were driving me to pursue the job. Why did I really want to work at Goldman Sachs? It certainly was not the content of the job itself. I did not have the slightest desire to pore over spreadsheets to understand how one company could acquire another. What drove me instead was a series of mental images: my mother smiling on the phone when she heard the news; my father proudly proclaiming to his friends, "Ravi has made it big in America!"; the postman delivering a money-order from my first paycheck to my parents; being in a position to loan Peter money instead of vice versa; finally being able to afford a superior piano and to play soulful jazz in a sparingly furnished, elegant apartment in San Francisco.

I suddenly became aware of John Schilling peering at me, blinking from behind his glasses, waiting for an answer. I shook off my musings, looked at him in the eye and let it rip. "I want to work for Goldman Sachs because your organization is a market leader in every sense of the word. In the areas of client satisfaction, financial performance and market reputation you consistently outperform your peer companies. You have long been the gold standard in investment banking and the excellence you demonstrate on a day-in/day-out basis is truly inspirational. I believe I can really learn and professionally grow in an organization such as yours. Additionally, I find the people I have met, yourself included, are collegial, warm, and friendly. So not only would I learn from the best minds in the business, I'll have fun doing it. This would truly be a dream job."

John Schilling smiled and held out his hand. From the warmth of his grip I knew I had made it through to the second round of interviews.

• • •

After the interview I ran back to my apartment, willing an e-mail from Carol to be sitting in my inbox. It was.

Ravi,
The bad news is that they cannot immediately locate the newspaper editions in the second half of 1919. These things

are not really well catalogued and it can be hit or miss, and this one's a miss. The good news is that our law library has access to the transcripts of the conversations between Vijay Sahni and Judge Taylor. You can request these transcripts at the law school. The case is Sahni vs. The People of New Jersey. There are twelve separate documents and the catalogue numbers are A28824–35.

Let me know what you find out!

Best,

Carol

So I would see the actual transcripts of Bauji talking to this judge!

The document custodian at the law school library was as unhelpful as Carol was helpful. It turned out that each transcript of the conversations was stored as a separate legal document and that I would have to request each of them individually.

"But they are all a part of the same case, can't I just get them all at once?"

"They have different classification numbers, sir; you will have to check them out one at a time, and you cannot check out more than one document a week."

"What? There are eight documents, so it'll be two months before I can read them all?"

"I am sorry to say that you have no choice in the matter," he said, not seeming sorry at all.

"How can that be? What do the law students do when they need more than one case at a time?" I asked.

"Usually there is no problem, but the age of these documents is the issue—they have special rules to protect these older documents."

"How about if I read them in front of you?"

"I am not going to stand here and watch you read, sir," he said raising his voice for the first time. "That is not in my job description."

I appeared to have no other choice so I just asked for the first transcript.

"I'll be back in about five minutes," he said, glad to have prevailed.

Twenty-two minutes later he handed me about fifteen typewritten sheets. I found an empty spot to sit and began to read Bauji's words.

• • •

April 21, 1919

Vijay Sahni vs. People of New Jersey

Official Court Document

Judge Taylor: Hello Mr. Sahni, I'm Judge John Taylor

Vijay Sahni: Welcome to my cell, judge.

JT: Thank you. Mr. Hanks over here is the court recorder. He will transcribe our conversation. As for me, you probably know why I'm here; the sheriff tells me you have been reading the local newspapers.

VS: Yes, I've been reading what the newspapers have to say, but that is never the complete truth, is it?

JT: Well, they are not too far wrong. Governor Williams has appointed me to determine whether the state should prosecute you for the crime of blasphemy or if you should be released immediately. I am helping the governor to decide this difficult issue.

VS: Yes, that is what I have read, but frankly speaking I have got the feeling that this arrangement may be a way of mollifying the big city press. They have consistently and vigorously expressed the opinion that my arrest is a travesty and that to try a person for blasphemy in this day and age is wrong. Apparently the governor is banking on the fact that your good name can cool things down in New York without letting down his constituents in Morisette. And it seems to be working, at least here in Morisette, for most of the local newspapers have welcomed your appointment even though they want the trial to go ahead. They feel you are a man who will do right by them, and that cannot be good for me.

JT: Mr. Sahni, it is not often that I need to say this, but no one has ever questioned my integrity. Let me correct you. I think the only reason the local papers have not criticized my appointment is because they feel that I will do what is right. What I think is right may not necessarily be in agreement with their views on the matter. I can assure you that this process will be fair and just.

VS: How can you guarantee that? Are you not merely an advisor to the governor?

JT: One of the conditions I imposed before accepting the assignment is that the governor abide by my recommendations whatever they may be.

VS: I do not believe you. Why should I go along with any of this?

JT: It is in everyone's interest, including yours, Mr. Sahni, that this trial takes place only if absolutely necessary. If you refused to work with me, your trial date would be at least three months from today. For that duration you will remain in this cell regardless. I will be finished with my investigation well before that. So if I decide against you, you have nothing to lose; you will go to trial as planned. If, on the other hand, I decide for you, you have everything to gain; in fact you could be released in as little as two weeks.

VS: I see.

JT: Let me be explicit on how I want to go about this: if I think that you said what you said out of malice and genuine contempt for Christianity, then I will recommend that a trial take place posthaste. If, on the other hand, I find that what you said was an impulsive response to provocation, or you had some other justifiable reason for saying what you did, then you will be set free.

VS: [unintelligible]

JT: Listen, Mr. Sahni, from what I can gather, you said what you said in the heat of the moment. I can totally and . . . [JT interrupted]

VS: You have it wrong, Judge Taylor. I may have said some things in the heat of the moment but there is nothing that I would recant or retract. I could have said it differently, but I stand by every word of what I said.

JT: I find it difficult to believe that any sane man could espouse the views attributed to you, but I don't want to prejudge the issue. This is what we need to discuss. If I can understand why you said some of these things, if the provocation was what spurred you on, then we could safely dispense with the trial.

VS: You really don't understand what I'm saying. This is exactly what I suspected would happen. Your own beliefs make it impossible for you to even comprehend that what I said may be my considered and rational judgment. My statement doesn't need justification; I have arrived at these conclusions after much thought and I stand firm on them.

JT: What makes you so sure of yourself? You are going against the great wisdom of the ages. When you blaspheme against Christianity, you question the very basis on which millions in this country base their lives. This nation's constitution itself is based on Christian values. You appear to be an intelligent man. I can hardly believe you said what you did, and what is worse, you appear to be standing by it.

VS: Judge, I stand by what I said at the town square. I think every free man has the right to think for himself and reach his own conclusions about these matters.

JT: I fail to understand your motivations. You are in a land where you have the freedom to live and think as you deem fit. You even have the freedom to exercise your own faith, whatever its nature, without anyone bothering you. Yet, you are the one bothering people and you persist in being stubborn about issues that are of paramount importance to so many people.

VS: There is no democracy in the realm of ideas. Just because so many people believe something or live in a certain manner is not reason enough for me to concede they are right. And your own laws allegedly give me the freedom to speak about such matters without fear of reprisal or arrest. My confinement violates some of the most basic principles of your constitution.

JT: Certainly, there is a legal, First Amendment issue involved here. But let's forget about legalistic arguments for a minute. I'm not here to try a case; I'm only here to decide if you have, by my lights, committed blasphemy.

VS: By your lights. What does that mean?

JT: Meaning in my opinion.

VS: I see. Look, judge, all the facts are clear. I'm not disputing the facts. I've already told you that I stand by what I've said. The ideas I expressed are inner beliefs of mine. I do believe that most of

Christianity, indeed most of religion, is quite . . . misguided. I did not say these things in the town square merely out of provocation. So what do you and I have to do together? You know what I said, and you must decide if it constitutes what you call blasphemy. It is quite simple isn't it?

JT: It is not quite that simple, Mr. Sahni. Actually I'm mostly interested in your motivations. Now, it is true that if this were a common court trial then your motives would not matter. If you were to steal a loaf of bread, you would be guilty, and there is no question about delving into motivations. But this is not a trial and the crime you are accused of is not at all commonplace. As a matter of fact the last blasphemy trial that was conducted in New Jersey was in 1885, over 30 years ago. So frankly, we're not quite accustomed to dealing with this type of situation. When I agreed to come talk to you, my chief aim was to understand you as a person, to understand what made you say the . . . extraordinary things you said; I wanted to understand your motivations because they will help me decide whether your case should go to trial or you should be set free.

VS: I'm not quite sure I understand you. What kind of motivations would be acceptable in your eyes?

JT: Mr. Sahni you must understand that this country's laws do not persecute people for the opinions they hold. Your motivations only become the State's business if it seems that you maliciously sought to inflame the passions of our populace.

VS: You have not answered my question. You yourself said that my motivations would help you decide my guilt or innocence. So I am asking you, Mr. Taylor, what sort of motivations would render me innocent in your opinion?

JT: Ordinarily I would refrain from answering that question, because you could then pretend to have those motivations. But your actions do not paint you to be such a man, so I will be open with you. A motivation that I would find acceptable would be that according to your religion, which I understand to be Hinduism—did I say that correctly?

VS: Hinduism, yes.

JT: Yes. If, for example, according to Hinduism, the path of Christ itself is incorrect, and you, perhaps in a fit of passion, started to proselytize in some fashion, that could be speech that is protected under the First Amendment. If, however, it were the case that your only motive was to blaspheme the Christian religion due to some aberrant personal philosophy, then I would be forced to recommend a trial and would personally support a conviction.

VS: But I am not a Hindu. In fact, I do not believe in God.

JT: You do not believe in any kind of God?

VS: No.

JT: Not even in the little idol gods they have in India?

VS: No.

JT: So you are an atheist?

VS: Yes I am.

JT: But I read that you worshipped the sun every morning.

VS: The newspapers got that wrong. I did go out for a walk every morning, and sometimes I did a few stretching exercises, but never any worshipping.

JT: I confess I am surprised. You being from India, I thought you would belong to some mystic religion or another.

VS: Most of my countrymen are intensely religious. But I am not.

JT: Are your parents atheists as well?

VS: No, they are not. As a matter of fact my mother is quite religious.

JT: So, if I may ask, what made you an atheist?

VS: It's a long story, but the turning point was brought about by mathematics.

JT: Mathematics? What do you mean?

VS: There is no simple explanation, judge. Mathematics provides a different way of comprehending the universe. Its methods rely on reason, not mysticism. Facts are established to be true only if they are proved; nothing is accepted on faith or authority. Its theorems constitute a real set of enduring truths, not made-up fantasies. Mathematics is certain, it is clean, it is fair, and it is just. Its methods have been shown to yield rich results in many branches of knowledge. Indeed, I have used mathematical rigor to examine all aspects of life, and in my studies, God has come up short.

JT: God has come up short? I must say that I find your statements somewhat childish.

VS: [laughter]

JT: Stop smirking, sir. I came here groping with subtle ethical and moral questions, weighing out freedom of speech versus the blasphemy law. Despite your expressing views that I personally disagree with, I came to you with an open heart, listening ears, and a nonjudgmental frame of mind. Instead of talking to me honestly, you have been smug, somewhat arrogant, superior, and haughty, and now you give me some malformed thought about not believing in God because of mathematics!

VS: Why do you think that is rubbish? Do you know enough about mathematics to be sure that what I say is rubbish?

JT: Surely you understand that there is no contradiction between logic and faith. At any rate, I am not here to discuss the philosophical implications of mathematics with you. I came here to ensure that you are treated as fairly as possible. I did not want the townsfolk's rising passions on this issue to interfere with justice and due process. I wanted to go the extra mile due to the special circumstances of this case. But now I see that the majesty and fairness of our justice system may be wasted on you.

VS: Then go on your way, judge. You do not need to feel guilty anymore. I see that you will be able to convince yourself that I deserve the jail sentence that your Morisette jury will surely award me. Perhaps that's why you came here in the first place.

JT: How wrong you are! I did not come here to assuage any guilt you might think I have over your arrest. Astonishing as this may seem to you, I came here hoping to do the right thing, indeed the Christian thing. The governor asked me to provide my opinion because he himself is not sure if there should be a trial. We understand that a trial would not be easy because there are almost no credible precedents to draw upon. The point/counterpoint of the state Blasphemy Law and the First Amendment has never been adequately debated, much less defined. From a purely legal perspective this case could be endlessly appealed whichever way the jury decides. So it would really be best if we didn't have to have a trial at all. If I could find enough cause to release you, then we would end this entire affair quickly. However, if I find that your actions may indeed have violated the blasphemy law in our state, then I will recommend that your case go to trial, however undesirable that option may be. You must understand that what you have said has deeply hurt and offended many, and the citizens of this county are rightfully demanding due process.

VS: On the contrary, I think that you came here wanting me to fit into a certain mold. Perhaps you wanted me to be the heathen who clings to his delusions about Christianity. Or perhaps you wanted me to be an emotional hothead and claim that I said what I did out of anger. In either case, you could complete your investigation and report that despite everything I said, my heart is in the right place after all. You would have me apologize to the citizens of Morisette, perhaps have me retract my statements, and thankfully avoid the messiness of a trial. That way the governor could give satisfaction to the people of Morisette without being labeled a backward religious zealot by the big city press. This is what the governor is hoping for, and this is why he has sent you to me. But I am not fitting into your mold at all. Instead of uttering comfortable platitudes, I am threatening to use rationality and logic to challenge your most cherished Christian beliefs. And this is what is making you uncomfortable. Look at you; your face is getting quite red.

JT: Sir, usually I am not an irritable man. People have even described me as patient and calm to a fault. But I must confess, you irritate me enormously. You appear sure that your thinking is superior to anyone else's. You have a tendency to boorishly lecture on, confidently assuming that you can see things that others cannot see. For example, you are sitting there, with your chin lifted upwards, completely certain that you are making me uncomfortable because I am somehow worried that you can challenge the validity of Christianity by your mathematics. I find that notion to be preposterous, not worrisome.

VS: Judge, I regret that I offended you; that was not my aim. But I really do believe that the methods of mathematics equip a person with better ways to look at life. They can make a man ask better questions and demand better answers about everything, including morality and ethics.

JT: But the laws of ethics are God-given, Mr. Sahni. They live in the Commandments. There is no such thing as demanding better answers about ethics, because God hands those laws to us. They are meant to be accepted and followed, not debated.

VS: No, I think that ethics are relative. Societies must decide what values they want their people to live by, and these values can vary from society to society.

JT: That is a perverse thought. I would not even have imagined that men were capable of it. But apparently I would have been wrong. Is murder OK in your society, Mr. Sahni?

VS: Of course not. Our worldviews are perhaps too far apart for us to even start a meaningful conversation. And your stubborn attachment to the Christian imperatives would quickly end the conversation even if we somehow got started.

JT: I am not being stubborn about anything, sir. If you detect an impatience and irritation in my voice, it is just because I cannot maintain a façade of politeness and understanding when you are saying things that make no sense at all.

VS: For example?

JT: For example, you are saying that you became an atheist because of your mathematical studies. "God came up short" was the expression you used. Now let me ask you, sir, what can mathematics say about God? What can logic tell you about faith? And then you tell me that you reckon that morality and ethics are subject to the conveniences of time and place. This conclusion, too, you appear to draw from your mathematical framework. As if the study of numbers and figures could tell you whether or not it is OK to steal! You are allegedly a man of letters, Mr. Sahni, but you are yet to show me that side of your character.

[Note from Court Reporter: Prisoner requested, and was granted, a five-minute break.]

VS: Judge, may I tell you a story?

JT: What about?

VS: It's story that may begin to answer your questions.

JT: Why the sudden change of heart? I thought our worldviews were too far apart.

VS: They may well be too far apart. But you asked me exactly the right question when you asked me what mathematics could say about God. It was the logical question given the context of our conversation. It is a question that deserves an honest answer. As long as there is logic and honesty on both sides there is always hope for progress. So may I tell you this story?

JT: Well . . . . Okay, go ahead.

VS: This happened when I was 12 years old. It was in India, in Lahore, the town of my birth. For weeks we had been hearing about what everyone was calling an incarnation of the Hindoo God Shiva. Apparently there was a little girl who under her own head had the body of a snake. Her head was perfectly normal, but instead of resting on a girl's body with arms and legs, it sat on the body of a snake. At first we could scarcely believe that this was possible, but neighbor after neighbor reported seeing the girl with their own eyes. Our whole town was completely enthralled by the phenomenon. As days went by, we began to hear that the girl could make any wish come true, that she could appear and disappear at will, that she could morph into other strange creatures, and that she could heal anyone's disease. It seemed that every day there was another miraculous story about this strange girl. Meanwhile, the girl's parents became celebrities in their own right. They had been in our town for only six weeks, but people showered them with love and generosity. One day my mother decided that it was time for us to go and pay our respects. When we got there, I remember thinking that it was the most number of people I had ever seen gathered in one place. There was a long line of devotees that snaked twice around the house. We waited for what must have been two hours, though at the time it seemed much longer. At the door, we were asked to pay the handsome sum of one rupee for the privilege of seeing the reincarnation of Lord Shiva. Inside the house there were cries of horror and delight as people went into the inner room and saw the snake-girl for the first time. Finally, it was our turn. My mother and I went inside. It was dark and cool inside the room. The creature sat in the center of the room on a platform covered by a white cloth. As advertised, there was a girl's head emanating from the body of a black snake. The snake was coiled up at the bottom, but rose up like a tree trunk, ending in the girl's head. The snake portion was completely motionless, but the girl's head seemed quite ordinary. She blinked completely naturally, and once, looking directly at me, she let out a slow yawn.

JT: There really was a girl's head on a snake's body?

VS: That is what it seemed like. There were flowers placed in a circle around the snake-girl's body. There were two cups of milk: one in front of her and one behind her. For some reason, it was the cup of milk that caught my eye. First, it seemed odd to me that she would need two cups of milk at the same time, and second, I couldn't understand why there was one cup in front of her, and the other symmetrically behind her. I had been reading my brother's school Geometry book, and one of the sections I had recently finished was on symmetry. Who knows why or how, but standing in that room, looking at that cup of milk, I had an Aha moment. I realized that the cup in the back was an exact mirror image of the one in front! Could it be that there was a well-placed mirror behind the snake's trunk, covering up the girl's body? Before my mother could react, I threw a pebble towards the snake's body. Sure enough, it bounced back. And then suddenly there were two pebbles!

JT: The real one and the reflection?

VS: Exactly.

JT: That is interesting. So the girl's body was behind the mirror and they used symmetry and reflection to bring the illusion to life.

VS: Yes, but it was not presented as an illusion. The girl was presented as God itself. When they realized I had seen through them, the girl's parents got me out of the house very quickly. At first my mother was angry, but she realized that something was amiss. I tried to explain that it was a hoax, but she never completely believed me.

JT: But you told everyone what was going on?

VS: Oh yes I did. I told everybody I could. Some people believed me, but most did not. They couldn't quite resist the evidence of their own eyes, especially when they were told that they were looking at the Almighty itself.

JT: So what happened?

VS: Nothing happened! The snake-girl stayed in the house and hundreds paid to see her every day. After a few weeks, when attendance began to decline a little, she and her parents moved on to another town, probably to repeat their trick. What amazed me was that life in our town continued fairly normally. In the following months people would sometimes talk about the snake-girl, but there was no uproar of curiosity, no investigations about how she came to be. When I told people the truth, many chose to completely ignore me and continued to believe in the miracle they had witnessed. They told me to be careful about what I said, lest I anger Lord Shiva himself. That didn't make any sense to me either. How could Lord Shiva be angry with me for spotting a trick? I tried to talk things through with people, but for the most part they did not listen. It was my first exposure to religious faith.

JT: Wait a minute. That was not religious faith, that was pure superstition. But pardon the interruption. Please continue.

VS: There isn't much more to the story, but it brought about some important changes within me. I was disillusioned with adults around me, but nevertheless, I had come out with the secret weapon of analysis on my side. I only believed things that were so clear in my own mind so as to exclude any possible doubt. For some years, I thought I was unique in doing this until I began to formally study mathematics. In mathematics, rigor is not merely suggested; it is required. Everything observed in numbers or nature must be understood using reason and reason alone. Mathematicians do this and are amazingly successful in explaining how things work. I had used the mathematical concept of symmetry to deduce that there was a mirror in front of the girl's body, and since then I have used mathematics, or the methods of mathematics, to understand everything I see around me.

JT: That is a very interesting story Mr. Sahni. Pardon me, but would you like to try one of these cigars?

VS: No thank you, I do not smoke.

JT: Then forgive me for lighting up. Yes, that is an interesting story indeed. You know, Mr. Sahni, I wish you had received proper guidance about religion at that critical juncture in your life. What you witnessed was empty paganism, not real religion. A girl's head on a snake's body—how childish, how stupid! For people to worship this apparition would be funny if it weren't so tragic.

VS: But I think that all of religion has a blind, unquestioning faith at its root. You Christians, for example, believe that a virgin gave birth, that a man could walk on water, and that he was somehow reborn after he was already dead. Yes, it is not a girl with a snake's body, but the underlying method of thinking is the same.

JT: No, not at all. The Lord Jesus Christ was not an ordinary mortal. The laws that apply to you and me did not apply to him. Surely this is not that extraordinary an idea; I see it everywhere. For example, I know enough mathematics to realize that the laws that apply to finite quantities may not be applied to infinite ones. It's the same with Christ. It is insane to compare his life and deeds to this . . . snake-girl. She was just a con artist sucking money from simple people; Jesus was everything that is noble and sublime.

VS: Judge, please do not misunderstand me. I am not claiming that Jesus was interested in cheating people out of their money. I don't know that; in fact, I really don't know anything about Jesus. What I am saying is that the type of unquestioning faith that made people in my town believe that the snake-girl was a personification of God is similar in nature to the faith that makes current-day Christians believe that Jesus was a miracle worker. Not because Jesus was trying to trick the Israelite and pagan people in some fashion, but because they, like people everywhere, did not apply the rigor of analytical thought to the world around them. They were driven by forces other than reason.

JT: Listen to what you are saying Mr. Sahni! According to you, the faith of the Apostles was mistaken, the saints who spread the good news were mistaken, the believers who preferred to be fed to lions rather than renounce their faith were mistaken, and millions of believers all over the world have been mistaken as well. But you are apparently not mistaken. You have somehow been blessed with this great tool of reason that enables you to pass judgment on us misguided believers. Why, pray tell, have you been so uniquely chosen?

VS: I have not been uniquely chosen. All I have done is read mathematics and understand the notion of proof, and I have applied this notion to everything around me. I do not accept anything without proof. I demanded proof that the snake-girl was real, I demand proof that Christ did all the things you Christians say he did, and I demand proof that God exists.

JT: Faith provides all the proof I need

VS: Judge, that does not constitute proof! What is the guarantee that the Bible's word is true?

JT: What is the guarantee that anything is true?

VS: The only guarantee, my dear judge, is this proof I speak of! Proof provides truth that is utterly beyond doubt. May I demonstrate, using an example?

JT: Demonstrate what?

VS: Demonstrate a truth that is utterly beyond doubt.

JT: I think the existence of God is utterly beyond doubt.

VS: You believe that, but you cannot prove that in any rigorous way. Perhaps my example will explain what I mean. May I?

JT: OK, if you must.

VS: Are you familiar with the Pythagorean theorem?

JT: The Pythagorean theorem. Well, yes, I remember it from school. Let's see . . . I think it said that the sum of the squares of the two sides is equal to the square of the hypotenuse.

VS: Almost correct. The theorem holds true for right-angled triangles only. It states that in a right-angled triangle, the square on the side subtending the right angle is equal to the squares on the sides containing the right angle.

JT: OK. So what?

VS: Judge, let me ask you, do you believe the Pythagorean theorem?

JT: Believe it? Yes, of course. It's been known for hundreds of years.

VS: So what? The fact that it has been known for hundreds of years in no way guarantees its truth. People believed that the earth was the center of the universe, in fact religious texts explicitly stated this to be the case. Yet it turns out that it is not even the center of the solar system we live in.

JT: Religious texts are often metaphorical. Not everything is meant to be taken literally. But more broadly, you are correct. The historical acceptance of a statement as fact does not make it so.

VS: Precisely. Further, if you think about it, the theorem is profoundly surprising. We are so used to it that we take it as intuitively obvious, but if you look at it anew, it is a startling result. Let me draw this out for you.

[Note from Court Reporter: Court Reporter asked if said drawing was to be made a part of the transcript. Judge Taylor asked Defendant if there was soon to be a link between this geometric discussion and the blasphemy issue at hand. On receiving affirmative assurances from the Defendant, Judge Taylor approved the inclusion of geometric drawings in the transcript. Messrs. Buckley and Sons of Morisette Engineering Works rendered this drawing.]

VS: So here is what this theorem is really saying: if you take a right triangle and draw a square on each of its three sides, then the areas of the square obey the following relationship: Area of square A + Area of square B = Area of Square C. This, then, is the Pythagorean theorem. Does the picture make the statement clearer?

JT: Yes, it does.

VS: Good. Now, Judge, pretend no one had ever told you about the Pythagorean theorem. If you can, look at the picture with fresh eyes. I ask you, is this not an utterly unexpected result? I mean, there is no immediately intuitive reason for it, yet the relationship between the three squares on the sides of a right triangle always

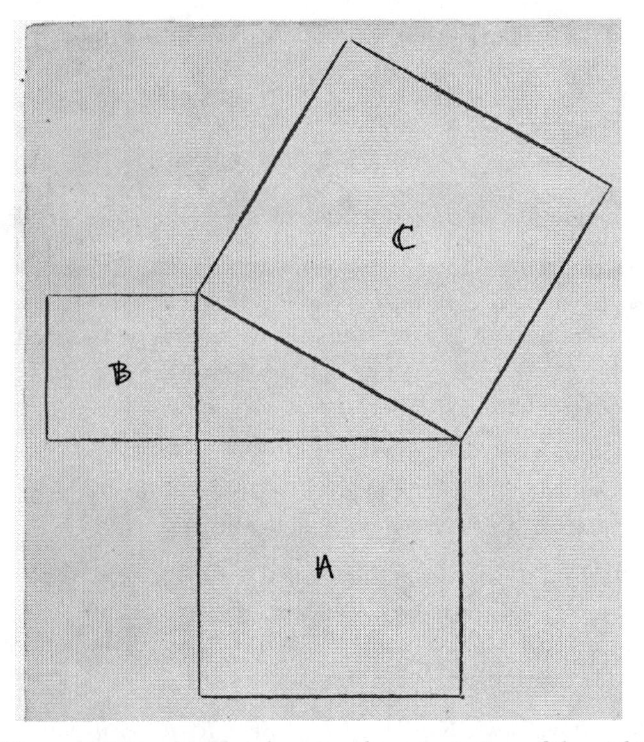

holds, no matter what the shape or the orientation of the right triangle. The relationship is precise and simply stated, yet it seems almost too precise, almost magical in its concise severity. It is hard to believe this result, much less be certain of its truth. Am I alone in this or do you see what I mean?

JT: No, you are right. It is indeed a surprising result, if you let yourself forget its familiarity.

VS: Now let me ask you again. Do you believe the Pythagorean theorem? Are you certain of its truth independent of the authority of anyone who has taught this to you?

JT: No, I am not certain of this result if I do not rely on the knowledge of others, Mr. Sahni. But then again, that is not too surprising. Most knowledge is obtained by standing on the shoulders of others before us.

VS: Judge, I am not saying that we should not benefit from the teachings of others. My point is merely that had I shown you this result for the first time today, you may not have believed it to be true for all right triangles everywhere.

JT: Fair enough.

VS: Thank you, Judge. Thank you for listening to me. Now my objective is to show you a method that will give you absolute certainty about the truth of the Pythagorean theorem—a result that, as we discussed, is surprising, and is certainly nontrivial.

JT: OK, but I'm not sure where certainty about the Pythagorean theorem is going to get us.

VS: It's going to show you the standards and the rigor I require to believe that something is true. And I believe that once you see the analytical methods of mathematics you will agree that it provides the only reliable path to human knowledge.

JT: That is truly a tall claim. Well, let us get started then. Let us see your method of guaranteeing certainty about the Pythagorean theorem.

VS: Good. Let me invite you to look at this triangle

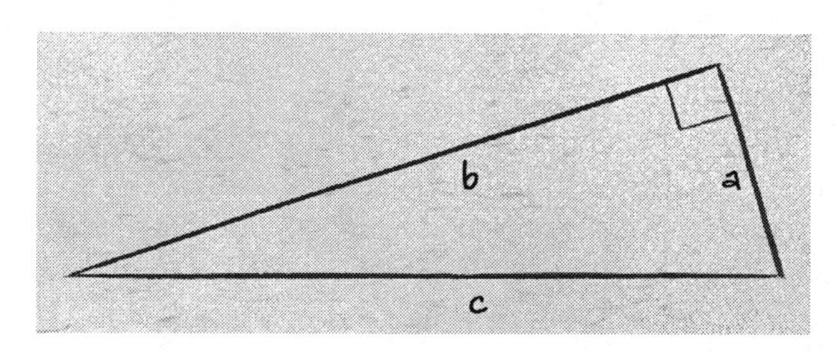

[Note from Court Reporter: Judge Taylor briefly examined the above figure.]

JT: This is a right triangle, just like the one you drew earlier. Only you've flipped it and stood it on its side. So the Pythagorean theorem should apply to this.

VS: Absolutely, Judge. How would the theorem read in this specific instance?

JT: You mean what would it say? That the square on side c is equal to the sum of the square on side a, and the square on side b, or $a^2 + b^2 = c^2$.

VS: Precisely! Your mind is quick, Judge! Now I invite you to contemplate this figure:

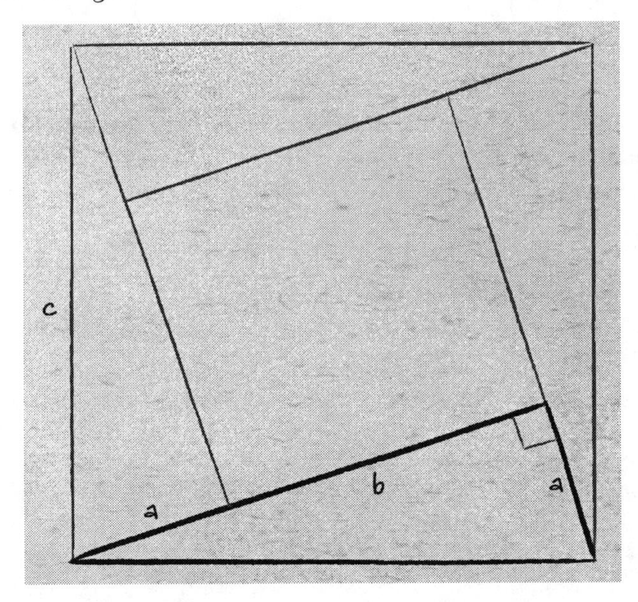

[Note from Court Reporter: Judge Taylor examined the above figure for approximately three minutes.]

JT: I see that you have drawn one of the triangles perpendicular to the b side of the original triangle, and another triangle along the a side of the original triangle. A fourth triangle easily fits there on top. Further, two squares result from the placement of the four triangles: the one outside whose length, as you indicate on the drawing, is c, and the one inside between the four triangles.

VS: That is quite right, Judge, and quite impressive. Are you mathematically trained?

JT: No. I am a lawyer by training. I have not studied mathematics since my school days. Why do you ask?

VS: Your analysis is very quick and very perceptive. You have analyzed this drawing more quickly than I would have expected from someone not familiar with the ways of mathematics.

JT: I have tried my hand at architecture and engineering drawing. And regardless of your apparently low opinion of our intellectual abilities here in Morisette, we do have quick minds that can compete with anyone, anywhere, and on any subject. Anyway, please proceed with your demonstration.

VS: From what I have seen of your fellow townspeople, Judge, your abilities are an exception rather than the rule. Regardless, I will proceed with the demonstration. You pointed out that there is an inner square between the triangles. What is the length of the sides of that square?

[Note from Court Reporter: Judge Taylor re-examined the above figure for approximately two minutes.]

JT: If you look at the triangle that adjoins the left side of the square, the complete length of that side of the triangle is b and the triangle on top has a length a above the square adjoining the triangle on the left. So the length of the sides of the inner square is the difference between b and a, or b – a.

VS: Absolutely correct. Now consider this second figure:

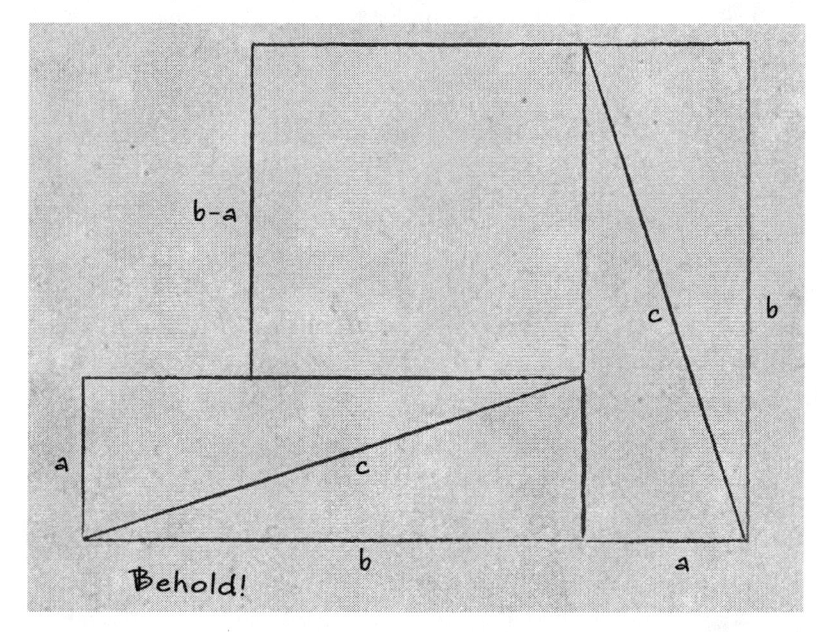

JT: This is the same drawing as the first one. All you have done is split up the triangles and rearranged them in a different way.

VS: Precisely! Would you agree then that the two drawings have the same area?

JT: Of course. The second is just a rearrangement of the first. Oh, I see what you're after! The first area is c^2 because it's the area of the square with each side of length c. The second drawing has two rectangles of sides a and b and a square of side $b - a$, so its area is $2ab + (b - a)^2$.

[Note from Court Reporter: Judge Taylor wrote out the following equation: $c^2 = 2ab + (b - a)^2$.]

Simplifying out the terms we get $c^2 = a^2 + b^2$, which was what you were after this whole time!

VS: Yes! Well done! You have just proved the Pythagorean theorem.

JT: Yes, that is a clever construction. Did you think it up yourself?

VS: No, Judge. It was demonstrated by the Indian mathematician Bhaskara in the 12th century. He simply drew the two figures side by side and wrote one word underneath the two drawings— "Behold!"

JT: I don't think it's quite that simple, but I grant you that is a nice piece of reasoning. So now what? What does this have to do with anything we were talking about?

VS: Yes, I am coming to that. Let me ask you, are you certain now about the truth of the Pythagorean theorem?

JT: Certain about it? Yes, I reckon so; the drawing forces it to be true. If the theorem was not true we could not construct the first drawing or decompose it the way we did in the second drawing. But the fact that we can implies that the result must be true. It is a direct consequence of us being able to construct the drawings, is it not?

VS: You are quite correct. We drew two figures: the area of the first was equal to c^2 and the area of the second was equal to $a^2 + b^2$. Since we clearly see that the two figures have equal areas, we may therefore be certain that $c^2 = a^2 + b^2$, which as you doubtless recall is exactly what we are trying to demonstrate. Notice that our certainty does not rest upon faith or belief in a higher authority in any way whatsoever. Instead it rests upon logic and clarity of thought. And this, Judge, is my key point: mathematics provides certain truth. It is the gold standard of human thought; its truths are eternal, absolute, and universal. You see this fact we just proved will always be true, no matter what. And it will be true even on Mars. If we were ever to meet the Martians we may not agree on much, but we would agree on the Pythagorean theorem.

JT: But this is one isolated piece of reasoning. What can it tell you about truly important things? You can have demonstrations about numbers and geometric figures, but what good are these? What do they tell you about what life means, how it should be lived, and why we are here—the truly important questions?

VS: Judge, the world around us operates according to mathematical principles. The methods of mathematics have succeeded in disciplines ranging from physics and mechanics to chemistry and the life sciences. Increasingly there is an exciting shift to bring the methods of mathematics even to softer subjects such as sociology and psychology. I contend that all human knowledge will one day be mathematized. Mathematics and its scientific cousins will have plenty to say about how societies evolve and how humans behave, and they will provide the keys for more learning and more understanding about what is really true. The religions that people today hold in such reverence will one day be exposed to be merely fairy tales with very little to say about how life works. I spoke in the park the way I did because I was reacting to the ignorance of people whose lives are unexamined and founded on the tenets of blind faith and superstition. But I believe that this will change. One day all human problems, whether of science, law, or politics, will be worked out systematically, with logic and clarity, and will be based on reliable and certain facts, not the fancies of some writer of a so-called religious text.

JT: You have made your position clearer to me, Mr. Sahni. Let me say that I do not agree with your thinking, but I am well aware that my agreement or disagreement is quite irrelevant. As I said earlier, my only purpose is to determine if there was an acceptable motivation for your speech in the park that could perhaps be protected by the First Amendment. If I do find this to be the case I will recommend your release. Else, I will recommend that the county proceed to trial. Our conversation today has given me many new insights into your thought processes. Doubtless I will have more questions to ask you but I would like to stop this conversation for today. I will consider what you have said so far and will come and talk to you on Monday morning.

• • •

BHASKARA JOURNAL ENTRY, 1150 AD

My dearest daughter Lilavati is emerging from the sorrow of her husband's death. For months after his going she did nothing—she just sat on the steps of the verandah and stared at her lap, ignoring even the little number puzzles I would leave for her amusement every night before going to bed. But one day I saw that she had written the answer to a very difficult problem I had thought up the night before. She greeted me that dawn with a small smile that gave me renewed life.

Now she has blossomed anew, and I have only mathematics to thank. These days, she is amused by justifying geometric truths. Just today she had a very interesting diagram to show that the sum of the squares on a right angle must equal the square on the hypotenuse. Everyone knows this result, but Lilavati showed why the result must be true. She brought me her construction with a single word, "Behold!" underneath it. I stared at her picture for a few minutes and realized what she was getting at, and then we laughed together. I will include this construction in my book, which I think I will name after her: Lilavati.

I'm not sure I understand her compulsion to prove everything. I am more interested in finding new truths.

• • •

Nico started off the third week's class by summarizing what we had done so far. "In our previous classes we have seen that the infinite is penetrable. We have worked with infinite sums that converge, and others that diverge. The main point I want you to take away from our work so far is that the infinite is not some mystical concept, akin to, say, religion, that cannot be analyzed using the tools of our mind. We have looked at series that go on forever and decided with certainty whether these series converge or diverge, and if they converge we have been able to identify the value they converge to. The human capacity to be able to play with infinity and to begin to make sense of it represents one of the greatest achievements of our species."

I realized that Nico—more than a good mathematician—was a good communicator. Teaching mathematics, like teaching any art, requires the ability to inspire the student. Inspiration requires marketing, and marketing requires stirring communication.

"And we're just getting started," continued Nico. "We're now nearly ready to look at Cantor's theory of infinity which is truly one of the crown jewels of human thought. Cantor was able to show that not all infinities are of the same size and, even more surprisingly, that there are infinitely many different kinds of infinity! To fully appreciate what Cantor did we have to understand the exact meaning of rational numbers and real numbers, which is a delightful story in its own right."

Nico went to the blackboard and drew out a number line.

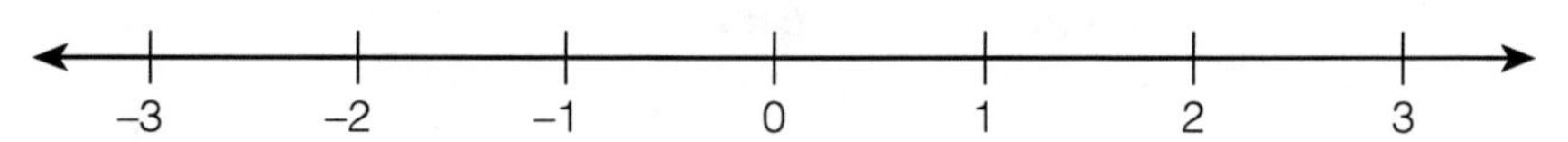

"We have here the set of integers, or whole numbers. Note that we've added the notion of direction here. The positive integers, also called natural numbers, begin with 1 and increase to the right. The negative integers begin with –1 and decrease to the left. The central spot is occupied by 0, the only number that is neither positive nor negative. Nothing very surprising here. Every number has a successor, and there is a gap of 1 between each number. Let me ask you, what's in the gaps?"

"The fractions," Peter said from the front.

"Indeed," said Nico. "The fractions are numbers of the form a/b, where a and b are integers and b is not 0. 3/4 is a fraction, as are 4/3, –2/7, and also 2, because 2 can be written as 2/1. Notice that some fractions are equal: 2/3 equals 4/6 equals 20/30. Another name for fractions is rational numbers. Rational numbers aren't any more or less rational than other numbers, it's just a name mathematicians have given to fractions."

Nico paused; he had a gleam in his eye, and I knew something interesting was coming up. "What's really interesting about the rationals is that they are endlessly close to each other," he said, "meaning that between any two rationals we always have another rational. For example, between 1/10 and 1/11, which are certainly quite close, we have 21/220. Anyone know how I got to 21/220?"

"Sure," said Peter. "You added them up and divided by 2."

"Absolutely," said Nico. "And the interesting thing is that I can do the same thing for any two rational numbers no matter how close they are. This is very different from the material world. I can take this piece of chalk and divide it into two, and at least theoretically continue the process until I get to the individual atoms. But there are no atomic mathematical intervals!

Between two rational numbers less than 1/trillion apart, there is still an infinity of other rationals. It is in this sense that the rationals are infinitely close to each other."

I tried to comprehend the immensity of the infinity of the rationals. Between 1/4 and 3/4 you have 1/2, and between 1/2 and 1/4 you have 3/8, and between 3/8 and 1/4 you have 3/16, and the process did truly go on forever. It seemed to me that this had to be a larger infinity than that of the integers in some sense, but I was not quite sure how one could go about comparing infinities.

"It would certainly seem that our number line is overcrowded with rationals. There are infinitely many rationals even in the smallest possible segment of the number line. Surely there is no room for anything else, is there?" Nico paused, enjoying the suspense he had created. "In one of the most momentous events in human history the Greeks found that there is indeed room for an entirely new class of numbers. They found that there are quantities that do not correspond to the ratio of two whole numbers. Previous to this discovery they believed that this was impossible—that every quantity in the universe was either a whole number or a ratio of two whole numbers, and the demonstrable existence of a different type of quantity was a shock to the system, akin perhaps to what the discovery of extraterrestrial life would be today."

Nico looked at his watch. "Let's pause for the cause. Please be back in no more than 10 minutes."

• • •

PYTHAGORAS' LETTER TO PHEREKYDES 513 BC

O VENERABLE TEACHER,

I am in possession of the news of your serious illness and despite my belief that when the time comes your soul will find a union with the divine, I find myself greatly saddened.

It has been so long since we sat together on our hillside in Samos with me drinking in your wisdom, and so much has happened in the intervening years, that I am not quite sure where to start in this message to you. Should I tell you about my travels, my work, my mistakes, or my friends?

"Tell me what is important," I hear you say from many stadia away in Delos, and as usual you are right.

What is important is what I have learned, so I will describe some of my learning. My greatest learning has to do with the mathematical nature of the universe. Let me describe to you the truth I have captured and how I came about it.

You taught me how to play the lyre, and ever since I left the home of my childhood I have practiced on it. Now I play effortlessly. My notes seem to give people happiness and sometimes they even seem to help the sick get better. None of this is unusual or surprising. But one evening when I had set aside the lyre and walked to the blacksmith shops in the town market I sat listening to the sounds their hammers made. Sometimes the sounds were harmonious and other times they were dissonant. I sat there listening for a long time and in an inspired moment, it came to me that if the ratio of the heaviest hammer to the lightest hammer is an integer, the sound produced is likely harmonious; otherwise it is not. Imagine my delight when I found this to be the case when I weighed the hammers. My delight was doubled when I realized that two strings also produce harmonious tones if their lengths are in ratios of whole numbers.

It was then that I began to understand that reality is mathematical in nature. My belief was strengthened when I studied astronomy, geometry, and the laws that govern moving objects. Everywhere numbers are the components of nature. If a perfect attunement (harmonia) of the high and the low can be attained by observing ratios of numbers, it is clear that other opposites may be similarly harmonized. The hot and the cold, the wet and the dry, may be united in a just blend (krasis). So yes, dear Pherekydes, I do believe that every quantity in the universe is representable by numbers and their ratios. The entire cosmos is a scale and a number! I consider this to be my greatest realization, one that I have passed on to many students.

Once I realized that *reality is number*, I proceeded to study it. I found truths about triangular numbers and perfect numbers. I found that the sum of the angles of a triangle is two right angles. And as I started to find these truths it came to me that it is necessary not just to state truths but to demonstrate them. If the cosmos is all number and scale, it is crucial to be certain of what we claim to be true about number and scale, and the only way to certainty is mathematical proof. I know that you have seen my methods in the demonstration of the result about the sum of squares on the sides of the right triangle. Some have taken to calling this theorem after my name, even though its truth was known for a thousand years before me. Perhaps it is because I have taken it from an observation about triangles to a demonstrable truth. I have applied similar methods to several other results.

So here again are my two great findings: One, that the nature of reality is based on number, and two, that one may find certainty about numbers by building mathematical proofs and demonstrations.

My current work is to develop a demonstrably true philosophical system. The truth of this system shall guarantee spiritual purity. Because everything is number, the divine philosophy, too, must be based on the methods of numbers. The members of my society, both the mathematikoi and the akousmatics have begun to make progress on the path to purity whose commandments shall be based on number-based demonstrations.

My fervent hope is that you will internalize what I have written. Even more glorious is the dream that you will provide me your perspective on my learning and worldview.

But I cannot leave you, dear Pherekydes, without telling you about a vexing argument that seems to put my two great learnings at odds with each other. There is no one else on earth I would trust with this information because it seems to provide a glimpse of reasoning that apparently gives birth to a creature so unnatural that it could not possibly exist, and yet I cannot find anything wrong with the demonstration. Please treat what I write below in the highest secrecy. If word leaks out about this, I fear that all the work we have done over many decades will come to naught, at least in the eyes of those who do not think too deeply about things.

One of the mathematikoi that belongs to my society is named Hippasus. Two moons ago we were on a boat headed to a nearby island when he drew this picture that I shall reproduce here:

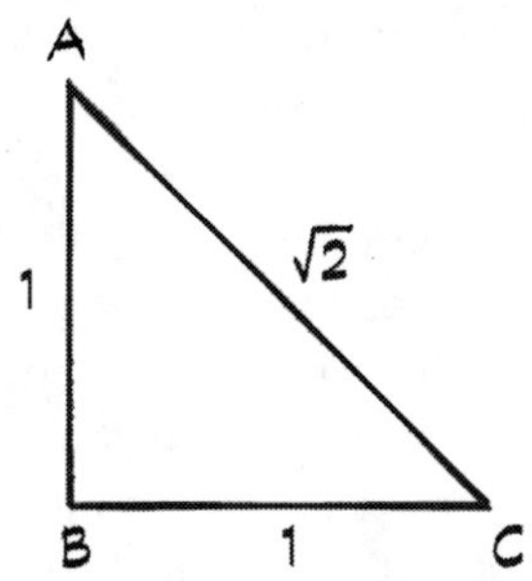

A simple triangle, is it not? Two sides AB and AC of length 1. By my own theorem, $AB^2 + BC^2 = AC^2$. Since $AB = 1$ and $BC = 1$

we get $1^2 + 1^2 = AC^2$. In other words, $2 = AC^2$ and therefore $\sqrt{2} = AC$.

Of all this there is no doubt. As I know you know, many have tried to estimate the value of $\sqrt{2}$. A close fraction is 14/10 ($(14/10)^2 = 1.96$); an even closer one is 141/100. Many of my students have made even better approximations—ratios whose square is less than an eyelash away from 2, but never exactly 2.

Here is where Hippasus made his extraordinary claim. He said that $\sqrt{2}$ is an irrational number, or a number that can never be represented as a ratio of 2 wholes. He arrogantly said that we could keep looking for two whole numbers the square of whose ratio is exactly 2, but we would never find such numbers.

Think of this, dear Pherekydes. What an extraordinary claim! Hippasus was saying that the length of the hypotenuse of his triangle could never be represented as a ratio of two whole numbers. This will destroy my belief that everything is number. It will create something that is beyond the reach of the whole numbers.

When he made the claim, many in the society were quite agitated. They told Hippasus that he knew not of what he spoke. But Hippasus countered with an argument that applied my own methods of mathematical demonstration to support his claims. In doing so, he used my second learning (that of pursuing certainty through demonstrations) against my first learning (that all is number). I will record his demonstration below. It is maddeningly simple and perversely beautiful in its way.

Hippasus says to assume that it is indeed possible to write $\sqrt{2}$ as a ratio of two whole numbers, say p and q. So $\sqrt{2} = p/q$. Now put p/q in its simplest form. By simplest form I mean cut out all the common factors and reduce it to a ratio that has no other common factors. For example, 6/8 and 3/4 are the same ratio, but 3/4 is the simplest form because 3 and 4 have no common divisors: $2 = p^2/q^2$, so $p^2 = 2\,q^2$.

In other words p^2 is even since it is a multiple of 2. Notice that the square of an odd number is odd ($3^2 = 9$, $5^2 = 25$, and so on). So p must be even, since if p were odd then p^2 would have been odd. Since p is even we may write $p = 2r$ for some whole number; it then follows that $p^2 = 2r \times 2r = 4r^2$. In other words, if p is even, then p^2 is a multiple of 4.

Having established that p is even, Hippasus asks about q. Since $p^2 = 2q^2$ and $p^2 = 4r^2$, we can say that $2q^2 = 4r^2$. Simplifying, we get $q^2 = 2r^2$. This means that q^2 is even and therefore q is even (again, because if q were odd, its square would have been odd as well).

So both p and q are even. But this is impossible since we had assumed that p/q was in simple form and they had no common factors. If p and q are even they will have 2 as a common factor. For example, 14/10 is really 7/5.

Therefore, Hippasus told that it is impossible for such a p and q to exist, for if they existed they could be reduced to their simplest forms. But we have shown here that such p and q must both be even and therefore are not in their simplest forms. A contradiction arises as soon as we assume that such p and q exist.

It pains me to say this, but I do believe that Hippasus' demonstration is unassailable. In structure it is similar to my demonstration showing that there is no largest prime. I too assumed that there is a largest prime and derived an assumption from that contradiction.

But if Hippasus is correct, then what length could the diagonal of a unit triangle possibly be? The whole numbers and their ratios occur everywhere in nature. You can have two ships and five boats and a ratio of 2/5 of ships to boats. But a quantity that is neither a whole number, nor a ratio? How can such a quantity exist? What does it mean?

As soon as I saw Hippasus' argument I besieged him not to talk about it outside the semicircle. I told him that we had finally shown the world that all nature was number, and his demonstration would prove that our worldview was perhaps incorrect and certainly incomplete without suggesting any better alternatives. Ordinary people would make fun of all the good work we had done over the years. But in his youth and pride, Hippasus refused to hold his demonstration in confidence. That night a few of the mathematikoi overcome by anger and worry over his stubbornness threw him into the sea. Not a swimmer, Hippasus quickly drowned. I feel bad about his death, but glad that the incommensurable hypotenuse shall remain within the society.

But I treasure knowledge even more than secrecy, which is why I have told you everything. I now face two alternatives regarding my future work: to attempt to develop a theory of the mysterious incommensurables, or to further develop my learnings in music, astronomy, and geometry, and to confirm that perhaps with one exception, all is indeed number. Any advice you may see fit to provide will be gratefully received and implemented.

I see this letter has taken the better part of this moonless night. I trust that you will receive it in good health. God permitting, I will visit you in Delos once the winter lessens its ferocity.

I still remain your humble student,

PYTHAGORAS

• • •

As soon as the class resumed, Nico plunged into the proof showing that $\sqrt{2}$ is irrational. He said that the proof was one of the most important in all of mathematics and was originally developed by a Pythagorean named Hippasus. Hippasus' proof—or at least Nico's retelling of it—was really so simple that when he finished sketching it out, I wasn't even aware that we had actually proven anything. The silence that followed showed I was not alone. Nico paused for a few minutes to let us mull it over.

It was Peter who broke the silence, "I'm not sure I understand what we have done."

Nico seemed to be expecting such a response. "Step back and examine the proof; in fact, you should try and do this with every proof you see or have to work out for yourself. First consider the structure of the proof—what we have here is the classical proof by contradiction. The Pythagoreans began by assuming that $\sqrt{2}$ can indeed be written as a fraction. Assuming this as a truth, you reach a contradiction through a series of logical steps, each of which follows from the previous one. If all the steps are correct and the result is an absurdity, then what we started with must be in error. We are left with no choice but to conclude that the very assumption that $\sqrt{2}$ is a fraction is in error. And in this context the actual contradiction comes from the fact that our logical series of steps ends up equating a square with a quantity that quite clearly cannot be a square."

He again waited for his words to sink in, and it began to make sense for me. All my mathematics teachers (other than Bauji and Nico) always seemed to evade this part of their responsibility. They had been content to merely write out a proof on the blackboard and carry on, seemingly without concern for what the proof meant and what it told us.

"But you should not stop here. Even when you have understood a proof, and I hope you have indeed understood this proof, ask yourself the next question, the obvious one, but as critical: So what? Or, why are we proving this? What is the point? What is the context? How does it relate to us? To answer these questions we have to step back a little. Let me show you—it's really quite delightful." Now there was excitement in Nico's voice.

"Let's look at fractions in a way we're more familiar with," Nico said walking up to the blackboard.

$$\frac{1}{2} = 0.5$$

$$\frac{11}{5} = 2.2$$

"In fact, we can write every fraction as a decimal number simply by dividing the numerator on top by the denominator below. The number we get is what we call a decimal number, as you must have studied at some point in school. But in the cases we have considered above, the division terminates within a finite number of steps. But this is not always the case. What is the decimal corresponding to 1/9?"

Nico waited for us to finish the division. It wasn't difficult, but while I was trying to figure out where this discussion might be leading, Peter put up his hand. He was always far more practical. He had been asked a question, he had figured out the answer, and he was going to spell it out.

"It is simply .1111, with the 1s going on forever."

"How do you know they go on forever?" Nico wanted to know.

"Well, each time you divide 10 by 9, you get a remainder of 1 and then you have to work out the same division with the same remainder again, so it's an unending process," said Peter.

So we encounter infinity again," said Nico. "What about the fraction 3/7?" Nico asked.

This time I worked out the division for myself. The answer I got was .428571 and then the entire block started repeating. Instead of just a single number an entire block was being repeated. So the answer was 0.428571428571428571. . . .

"When do you start getting the repetition of a block?" Nico seemed in no mood to help us out today.

As he waited for an answer I went over the division again. The decimal itself seemed to hold no answer. The fraction 1/9 had started repeating immediately, while in this case a block of 6 was being repeated. Then it struck me: Peter had already answered the question. It was not the decimal I should look to but the process of division itself. Obviously, as soon as we obtained a remainder that had already occurred earlier in the process, the numbers would start repeating. I blurted this out.

"That's just it," agreed Nico. "What Ravi is saying is that once you have done the division by 4, 2, 8, 7, 5, and then 1, you end up with the remainder of 3, so you are back to the original division of 3 by 7 and you will get the same set of numbers again. Now I am going to make a far stronger claim, but it follows from the arguments Peter and Ravi have sketched out for us. All fractions give us decimals that are terminating, such as 1/2, or they are nonterminating but repeating decimals, such as 1/9 and 3/7."

I was already back to my notebook. It seemed to me that the argument I had just sketched out worked for every fraction. If the fraction terminated, then, well, there was nothing to show. But if it did not, then I just had to look at the remainders again. In any division the remainder had to be less than what was dividing the number, so if the fraction was m/n, then there were only n − 1 possible remainders. If the decimal did not terminate after n − 1 steps, then a reminder that had occurred earlier in the division would occur again and the decimal would start repeating. Of course, I didn't work this out in this step by step fashion. The answer came to me all at once, and I somehow knew it was correct, another one of those Aha moments which had evidently made a reappearance in my life. In this case the Aha came with an associated insight. I realized that *any* fraction with 7 in the denominator would repeat after at most 6 steps. A fraction with 39 in the denominator would repeat after at most 38 steps. Claire, too, had come up with the same thought. She sketched out the argument on the blackboard. Only because she was standing next to Nico's tall frame did I realize that Claire was no more than 5′3″. Had someone asked me her height before that moment, I would have overshot the mark by at least three inches.

Nico was pleased with Claire's demonstration. "Isn't that beautiful? From a few simple examples and the power of abstract thought we have been able to deduce that fractions, or rational numbers, can only be decimals that terminate or repeat. Now here's the next natural question: We know that $\sqrt{2}$ is an irrational number—it cannot be represented as a fraction. What does its decimal expansion look like? Does it ever terminate? Or does it repeat?"

My mind was going a million miles a minute. I wanted Nico to stop for a second so I could work this out for myself. Unfortunately he was up against the clock and had to push ahead. "The decimal expansion of $\sqrt{2}$ is a different animal. It neither repeats nor terminates. If it did, it would be a fraction and the Pythagoreans proved that $\sqrt{2}$ is not a fraction."

"Wait a minute! You haven't proved that," protested Adin. "How do you know that if a decimal expansion terminates or repeats, it belongs to a fraction? All you've shown is that a fraction must terminate or repeat."

Nico grinned. "I was wondering if anyone would catch that," he said. "The proof is simple and it is possible to see the general from the specific. Let me show you," he said turning to the blackboard.

"Say f = 0.50000. . . . Multiplying both sides by 10, we get $10 \times f = 5$. Now dividing by 10, $f = 5/10 = 1/2$. So f is indeed a fraction," said Nico.

"Funny you called it f, for fraction!" laughed PK.

"A little prescience," acknowledged Nico, smiling. "Now let's examine a repeating decimal f = 0.3333, . . . $10 \times f = 3.3333$, . . . f = 0.3333. . . . Subtracting f from $10 \times f$ we get $9 \times f = 3$ or $f = 3/9 = 1/3$.

It was a neat little ploy, reminiscent of what Claire had done with the infinite series in the first lecture. Only this time it was legal. We were allowed to cancel the infinity of terms because we knew the underlying geometric series converged. I saw Adin scowl as he sensed the issue, and release as he saw the resolution.

"So now we can safely say that the decimal expansion of an irrational number cannot terminate or repeat. To be sure, we have here yet another encounter with the infinite. But keep in mind, there is a difference, a difference that took a genius like Cantor to first appreciate in all its significance. A repeating nonterminating decimal is basically finite information repeated infinitely often. After all, 3/7 was a decimal with 6 numbers repeated infinitely often. But $\sqrt{2}$ will be an infinitely long decimal since it doesn't terminate and it will, in some sense, have infinite information; after all, since it doesn't repeat, knowing one part of the decimal is not sufficient to determine what follows. This is the hallmark of an irrational."

Nico was interrupted by Adin, "So what is an irrational exactly? Is it a decimal expansion that does not repeat or terminate? That seems too imprecise. It is defining something by what it's not."

Nico nodded. "Recall the history of this, Adin. The Pythagoreans encountered $\sqrt{2}$ as the hypotenuse of a right-angled triangle that has both sides equal to 1. It occurred so simply and so naturally that it caused the first intellectual crisis in mathematics; it called into question the entire world view of the Pythagoreans, who believed that the world as made by God was mathematical. This world view was based on the integers, including ratios of integers, which you can now see is just another name for fractions. But they were shocked to find that the integers and their ratios were just not enough. They needed the irrationals, much as they were terrified by their unabashed infinity.

Nico was pacing as he talked, looking out the window. It seemed to me that he was back with the Greeks two thousand years ago and I felt sure that they would have loved to have him. A squirrel rustling the branch outside the window brought him back. He looked at Adin and continued, "Now as you point out, it is somewhat unsatisfactory to define an irrational as a decimal expansion that does not repeat or terminate. Neither is it helpful to say that an irrational is a real number that cannot be expressed as a fraction, for one may rightfully ask for the precise definition of a real number. One of the definitions modern mathematicians use is to think of a real number as a symbol for a certain sequence of nested rational intervals. Nested just means that each of the end points in an interval is contained in the preceding one, such that the length of the nth interval converges to zero as n increases. Hopefully you see that corresponding to each sequence of nested intervals there is exactly one point on the number line that is contained in all of them: if there was more than one point, you'd have some distance between the two points and eventually get an interval small enough to contain one and not the other, since the lengths of the intervals tend to zero. So there can only be one unique point corresponding to our sequence of nested intervals. This point by definition is called a real number."

Nico wanted to say more, but it was already ten past and some in the class were getting fidgety—it was time to go. But Nico couldn't let go; he was in love with these ideas.

"I want you to work out one last problem for yourself. Consider any number that is not a square, that is, a number other than 1, 4, 9, etc. Prove that the square root of such a number cannot be a fraction. Begin by showing that $\sqrt{3}$ is not a fraction and you'll see the general pattern. The proof is almost the same as what the Pythagoreans did for $\sqrt{2}$, but it does require some thought. If you believe this result about all nonsquare numbers, you automatically get the existence of a large number of irrational numbers: for example, $\sqrt{3}$, $\sqrt{5}$, $\sqrt{6}$, $\sqrt{7}$ are all irrationals."

"Wow, there's a whole family of these guys," said PK.

"Indeed. Now rational numbers and irrational numbers are together called real numbers. Any decimal expansion, whether or not it terminates or repeats, is a real number. Cantor made the extraordinary claim that the set of natural numbers is the same size as the set of rational numbers, but that the set of real numbers is larger than the set of rationals. He dared to claim that there were different kinds of infinity, and one infinity could be larger than another. Many thought he was mad to believe this, but in the end his ideas have prevailed. Infinity does indeed come in different sizes. That, ladies and gentlemen, is one of the gemstones of this course."

• • •

That evening I joined Claire for her daily run. I had thought that I was in decent shape and would have no trouble keeping up with her. But by the second mile, as we got to the Stanford foothills, it was apparent that she was slowing down so that I could keep up. "Were you able to find out about your grandfather?" she asked, as we took the first slope.

Between pants I told her about Bauji's speech, his arrest, the governor, and Bauji's quest for something certain to base his life on.

"He sounds a lot like Adin," said Claire.

I told her that the Bauji in the transcript did indeed sound like Adin, but he wasn't much like the Bauji I knew. I remembered my Bauji being at peace; he knew what he liked (mathematics and jazz) and he was content to stay within those things. Unlike Adin, he was not, or at least did not appear to be, obsessed with philosophical conundrums. I didn't even remember him talking about absolute certainty.

"Maybe he figured it out and didn't need to talk about it anymore," said Claire.

Maybe.

CHAPTER

1 2 3 4 5 6 7 8

I REMEMBER NOW HOW IMPATIENTLY I waited for Adin to finish reading the transcript. But he read with slow deliberation, stopping every once in a while to look away. Every time I thought he might be done he turned a few pages back and reread a section.

I had left him a message in the morning: "I need to talk to you. It's about something you can relate to and frankly you're the only one who might have a perspective on this kind of stuff."

When he called back, I was ready for him to fish for details, or at least inquire about the topic at hand, but "Where are you?" was all he asked. Fifteen minutes later he was in my apartment plunking down his backpack and unstrapping his bicycle helmet.

I told him everything I knew about the goings on in Morisette. The speech, the arrest, the judge, and the transcript. I told him that, like him, Bauji too had grappled with the issues of how we could ever be certain about anything. He listened intently, saying nothing, and once he started reading the transcript he did not look at me even once. I thought I would leave him alone and get some work done, but I couldn't concentrate. I kept waiting for his face to transmit reactions to what he was reading, but he stayed impassive the entire time. Finally, when I knew he had read each page at least once, I interrupted him.

"It is an interesting coincidence that after our conversation about certainty . . ."

"It is not a coincidence," he said, completely sure of this. "It is the most important question there is. That's why it keeps coming up."

His complete matter-of-factness made me feel uneasy. I had never spent much time thinking about if a thought is absolutely certain or not. My outlook at that time was utilitarian: If something worked, it was true, and that was that. If not, you tried something else. The issue of absolute certainty did not seem relevant. Yet, to Bauji and to Adin, it was the fundamental question. Was it a lack of seriousness on my part that I had not yet been troubled by this issue of absolute certainty, or was it a practical instinct that guided me toward problems that I could actually make some progress on? And was there progress to be made in pursuing certainty?

Adin must have sensed my ambivalence for he offered a defense to my unasked question. He spoke softly, punctuating his thoughts with slow blinks. "We are born into this world and we live our years according to certain beliefs. If the beliefs are wrong then we've squandered our time here. It is logical to ask how we can know if something is unassailably true. For if it is, then we have something to base a life on. But if we never find anything to be reliably true then we're just guessing, even with our most cherished values, values that people live their whole lives on. Your grandfather is saying here," he pointed to the transcript, "that mathematical methods provide a certain and reliable way of gathering knowledge, any knowledge. He is pointing out that the religious ideas of the judge are not true, not in the sense that the Pythagorean theorem is true."

"So you agree with him?"

"I don't know yet," he said, looking away. "Many philosophers, including Descartes, have written extensively about certainty, but their arguments have never satisfied me. Their arguments often seem circular or have unstated assumptions within them. But the idea of getting certainty through mathematics appears to be different. Before I saw Claire's infinity of primes theorem or the $\sqrt{2}$ irrationality theorem that Nico demonstrated, I wouldn't have known what your grandfather was even talking about. But these theorems were a revelation to me. I have no doubt about their truth and no amount of additional information could possibly change my mind."

"Those seem like two pretty good criteria for certainty."

"Yes," he said, surprised, and I think pleased, that I had come to the same conclusion as he evidently had. "So I have begun to glimpse how mathematical proof can supply certainty. But I'm not sure if the same methods can be applied to other forms of human knowledge."

I was not sure of that either. But it was not Adin's or my opinion that mattered, only Judge John Taylor's. And not being able to find out anything further on the Bauji front, I looked to immerse myself in the ideas of Georg Cantor.

• • •

GEORG CANTOR JOURNAL ENTRY, 1877

I end every supper by asking my wife if she has been pleased with me that day and if she continues to love me. I need to know that all is well with my family to be at peace and I need to be at peace to think about mathematics. Today, like on countless other occasions, she provided me the reassurance that I so ardently seek. "You are a huggable bear, Georg," she said. Many of my colleagues think of me as hot tempered and boorish (they tell me this to my face, I'm sure it's much worse behind my back). That there is someone in the world who refers to me as a "huggable bear" would be truly astounding to someone like Kronecker.

Damn that Kronecker! Even writing his name in this little diary of mine fills me with a certain queasiness. How he clings to his misguided view of mathematics! How he loves his self-imposed limitations! It's not that he lacks the imagination to understand my mathematics; on the contrary, he is a better mathematician than most. His trouble instead is that he refuses to venture out of the confines of the finite number. For him reality begins and ends with the finite number. He even denies the possibility of the existence of the nonfinite. And instead of treating his preference for the finite as a personal matter of taste, he chooses to relentlessly attack those of us who can behold the transfinite.

He has fought me long and hard, and the battle became personal long ago.

My logic seems so natural that I cannot find any mathematical reason for Kronecker to oppose me. I have gone over it a thousand times in my head, checking it and rechecking it. Reviewing my arguments has become a form of therapy for me. Today, perhaps because I'm too tired for my walkabouts (which always accompany my brooding), I'm moved to write down my thoughts.

Admittedly my conclusions defy intuition. For example, surely one would think that there are more rational numbers than Integers. After all, the rational numbers are dense. Between any two rationals there is another rational, no matter how close they are to one another. As such there is an infinity of rationals between 1/1000 and 1/1001. The whole numbers, on the other hand, are not packed. There is no integer between 1 and 2, for example.

So when I tell people, especially people like Kronecker, that there are as many Integers as there are rationals, they start out suspicious. But they should not stay suspicious. There is an excellent reason for my conclusions. Extraordinary claims merit extra-strong proofs, and I have them.

But let me backtrack a little. To compare Rationals to Integers, or indeed any one set to another set, one needs a method. I relied upon the simple method of one to one correspondence. If one wants to see if a set of five apples is greater or smaller than a set of seven oranges one might match one apple and one orange and set the pair aside. Then the process could be repeated four more times until one would have no apples left over, but there would still be two oranges, and this would lead to the conclusion that there are indeed more oranges than apples. This procedure allows us to conclude that there are more oranges than apples without having any notion of what the numbers 5 or 7 actually mean.

It seems silly to do this because everyone knows what 7 and 5 mean and everyone knows that 7 is greater than 5. But it's not so silly when you use this process to compare infinite sets. What if we could devise a method for comparing infinite sets on a paired basis?

Here's a really simple example. Say I want to compare all whole numbers to all even numbers. At first glance, it certainly appears that there should be fewer even numbers. Indeed you may hazard a guess that there are half as many even numbers as there are whole numbers. This guess turns out to be incorrect:

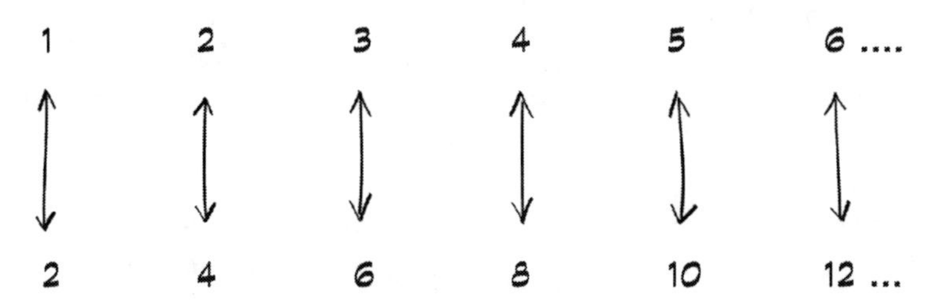

I've matched one to two and two to four and so on forever. Every whole number has a unique pair partner in the set of even numbers. 1001 is matched up with 2002; 5001 is matched up with 10,002. No matter which number you give me I can point to its unique partner in the set of even numbers and vice versa. I'll never have any elements of either set left over unmatched. Unlike the case where I still had oranges left over, even though I had matched all the apples, I'll never have any whole numbers (or even numbers) unmatched. This allows me to say that there are as many whole numbers as even numbers.

In a brilliant piece of insight many years before his time Galileo used a similar mapping to show that there are as many squares (1, 4, 9, 16, 25, . . .) as there are whole numbers.

One might begin to think that all infinities can be mapped equally into one another. This is not true! Had I found all infinities to be equal I would not have bothered to keep such careful notes, for there would have been nothing interesting to say. But fortunately or unfortunately it has fallen to me to show the world that with our minds we can behold the surprises of the transfinite.

Back to the question of pairing off the rationals and integers: it is not immediately apparent what such a mapping might look like. When I first glimpsed the answer I was quite awed by its simplicity. A rational number is just a fraction—or a ratio between two integers. Alternatively, it is also a terminating or repeating decimal, but it is easier to see the pairing with integers if you think of rationals as fractions. Some other day when I have some time I will construct a two-way mapping between the decimal expansion of rationals and integers, but for today let us lay out the rationals in the following grid:

1/1	2/1	3/1	4/1	5/1 ...
1/2	2/2	3/2	4/2	5/2 ...
1/3	2/3	3/3	4/3	5/3 ...
1/4	2/4	3/4	4/4	5/4 ...
1/5	2/5	3/5	4/5	5/5 ...
...	...	...	...	... ...

The first column of this array consists of all fractions which have 1 in the numerator; the second column has all fractions with 2 in the numerator. Similarly the first row contains all fractions with 1 in the denominator, and the second row has everything with 2 in the denominator. Notice that all fractions are contained in this infinite array. To find the fraction 17/33, for example, one merely has to move to the 17th column and the 33rd row.

Now the question I asked myself was whether I could traverse this array, one step at a time, and associate each positive fraction with a positive integer. The solution turned out to be delightfully simple:

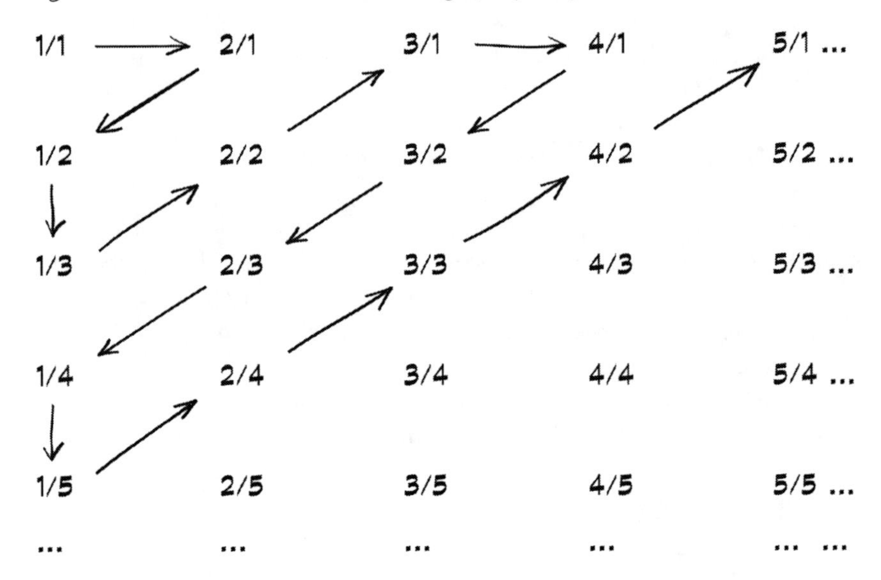

My path starts out at 1/1, then goes one step to the right, then diagonally to the left, then down, then diagonally up, and then again one step to the right, and so on forever. So we count off as follows: 1/1, 2/1, 1/2, 1/3, 2/2, 3/1, 4/1, 3/2, 2/3. . . .

Now, it is true that we will encounter fractions in different guises. For example, we have 1/1 as our first fraction and then 2/2 as the fifth, both of which are just different names for the same quantity. In such situations we just cross out the repeated occurrence and move on. In this way we have arranged all positive fractions in a row which we can then associate with the positive integers. And since they match up one for one, we may conclude that there are as many positive fractions as positive integers.

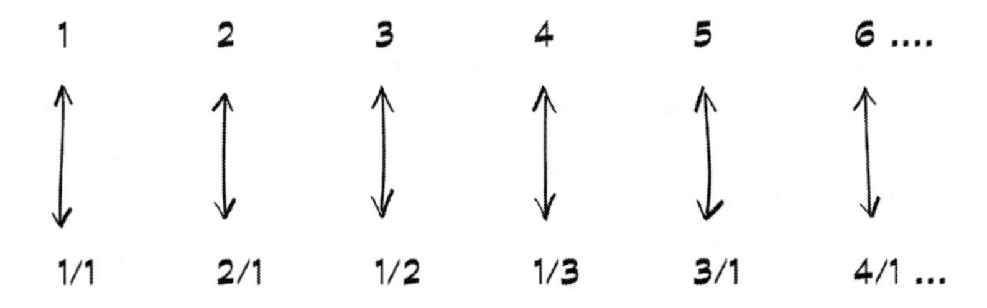

A nearly identical argument would allow us to include negative fractions to show that there are as many rationals as integers.

So there are as many rationals as there are integers! It seems to defy our native intuition, but we can prove it, so it must be true!

• • •

Nico started the following week's class with a thought experiment. He spoke slowly to allow our visuals to catch up to his words: "Imagine it's 1915 and you and your friend are in New Orleans on a vacation, out for a stroll. Getting a little lost you find yourself walking down a little lane somewhere near the Canal Street area. Also imagine that you are both white, and both musicians—you play in a string quartet in New York. As you're walking by a shady-looking basement bar you hear an odd noise coming from down below. It takes you a few seconds to realize that the noise is indeed intended to be music. Your friend covers his ears and wrinkles his nose in disgust. 'These guys are awful,' he says. But you're not so sure. Something about the music has grabbed you. It has a terse expressiveness and the notes are arranged in unexpected, new rhythms. 'Let's go down a minute', you tell your friend, not really giving him a chance to say no.

"Inside it is smoky and dark. As your eyes adjust you realize that everyone in the room besides you is black. You expect to have attracted more attention, but people mostly ignore you; they're listening to the musicians, moving to the beat. A handwritten sign on the back wall announces 'Earl's Jazz Orchestra'. You pull up a chair and start to listen."

"The band seems to defy every musical rule that you grew up with. You've never heard notes this high or this low. You've never heard timing this fast, or transitions this abrupt. You never knew that melody could be so superfluous, and that improvisations could be done so freely. The music has no business sounding this good, but it does."

In the next hour your life changes. Jazz, despite its unusual starting points and startling structure, explains music to you as you have never understood it before. Musical tendencies that you knew existed inside you have been freed and they in turn start linking with each other. You want to leap onto the stage and have a go yourself. Your friend, on the other hand, is unmoved, even a little repulsed. The music has made him question the very foundations of music and he doesn't like the feeling."

Despite my pre-occupation with Bauji's fate, I found myself engrossed in Nico's story. Jazz and mathematics were two of Bauji's great loves and I wondered how Nico would weave them together. He walked over to the large windows on the side and spoke looking at the trees, facing away from us. Many years later I learned that Nico always insisted on teaching in classrooms with windows.

"The reaction that the mathematicians of the world had to Cantor's creations was very similar to how our two musicians reacted to Jazz. While a few mathematicians bought in to the beauty and power of his ideas, many of the conservative elite ignored him or rejected him outright—he was treading on ground that was too unfamiliar, too arcane. For example, Cantor proved that the infinity of real numbers is greater than the infinity of integers. I personally believe this result to be one of the greatest achievements of humanity." He turned around and faced us as he said this, his head tilted upwards. "As you see, it is simply stated, and if you think about it for a minute, it is also truly startling. It says that there are layers of infinity, that infinity is not a monolithic, unanalyzable, and mysterious concept. On the contrary, humans can rationally and methodically get their arms around the infinite. And what's more, we can derive mathematically certain results about it."

He turned around and pointed to Cantor's poster, which he had unfurled and hung on the wall during the first class. "The genius of this man was that he saw outside the confines of the mathematical rules of his day. He dared to ask questions that others never did. He dared to draw conclusions that, however counterintuitive, had to be true. And he did all this despite facing vicious opposition from his colleagues. Respected mathematicians such as Kronecker were finitists in that they only believed in mathematical objects that could be constructed from natural numbers in a finite number of steps, so whatever arguments that Cantor made were dead on arrival. It didn't matter that his proofs were starkly beautiful, it didn't matter that the proofs really left no room for dissent, and it didn't matter that Cantor was no less a revolutionary than the inventors of Jazz or Impressionism or space travel—he was nevertheless relegated to the backwaters of mathematics, at least during his lifetime."

"I have a question," said Adin from the front, which as the semester wore on, would become his trademark preamble in Nico's classroom. "I can see how you show two infinite sets to be equal—you just find a one-to-one mapping from one set to another. This is what we did to match each whole to an even number, and also to map each fraction to a whole number. But how could you possibly show that such a mapping does not exist? I mean, just because you can't find one doesn't mean that a one-to-one mapping is not possible."

Nico liked the question. He was smiling before Adin was done. "To show that a mapping between two sets does not exist, you assume there is one and see if you get a contradiction."

This seemed in spirit like Claire's proof for showing that there is no largest prime. She had assumed that there was and then had logically proved the existence of a greater prime. I was sure that Nico had a similar structure in mind, but couldn't quite begin to grasp the specifics.

"Well, let's get into the proof," said Nico. "We have here one of my favorite theorems in all of mathematics, so for those of you who have not really internalized any of the stuff we have done before this, I ask you to pay particularly close attention. I consider this piece of thinking to be high art."

"Let's begin by recalling a few facts we discussed last week." Nico went to the blackboard and wrote:

1. A real number includes rational numbers and irrational numbers

2. A rational number is a terminating or repeating decimal (49, 3.6, 7.6358, 13.585858 . . . are examples of rational numbers)

3. An irrational number is a nonrepeating, nonterminating decimal (the square root of 2 or 1.414213562 . . . is an example of an irrational number)

"So a real number is really any decimal expansion you could write out?" asked PK.

"Absolutely. We discussed this last time."

"Are there such things as nonreal numbers?"

"Yes, there are such mathematical quantities. A pair of real numbers, for example, is certainly a valid mathematical quantity, and it is not a real number. We won't worry about these quantities here, but suffice it to say that any decimal expansion that could possibly exist is in fact a real number."

He went back to the blackboard and wrote out the objective:

We want to prove that there are more real numbers than integers.

"Now, obviously every integer is also a real number. But this fact alone does not tell us that there are more reals. As such, every integer is also a rational number, but as I hope you've read, integers and rationals can be placed into one-to-one correspondence with each other, implying that the two sets have an equal number of elements. So, it remains to be shown that there are more real numbers than integers. As I've said, we'll proceed by assuming that there *is* a one-to-one correspondence between the two sets. So say, for example, that the whole number 1 is matched up to the real number 0.8674339721 . . . and 2 is matched up to 0.5000000. . . . In this case 1 is matched up to a nonterminating, nonrepeating irrational and 2 is matched up to a terminating decimal, or rational. Both rationals and irrationals are reals so it's okay to include them in the reals column. For convenience we'll limit the real numbers to be greater than 0, but less than 1. This is not really a restriction because there are functions that map the interval [0, 1] into the entire number line in a one-to-one fashion. So all we need is a solution that applies to the reals between 0 and 1."

Nico drew out the example mapping as he spoke. After showing seven pairs he drew three dots to indicate that his list would go on forever. "There is nothing unique about this specific mapping. In fact, you will see that the same argument will work regardless of the mapping you begin with."

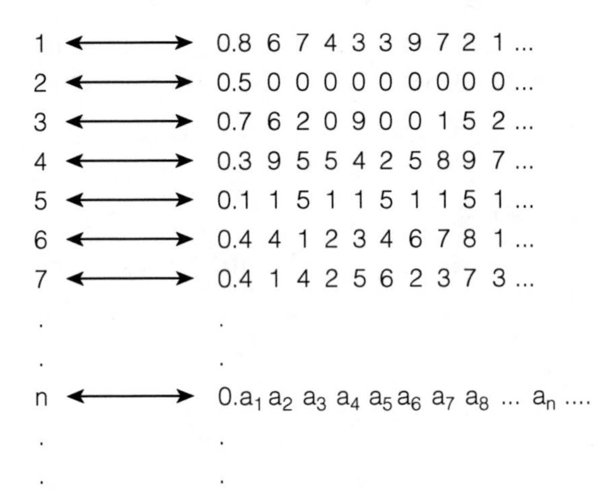

"Now, if this really were a one-to-one correspondence then every real number between 0 and 1 would appear somewhere in the right-hand column, matched with some integer on the left. This is where Cantor's genius shone through. He described the construction of a real number that could not possibly be anywhere in the right-hand column. He started with considering the digits in the diagonal, like this." Nico went on to circle the numbers down the diagonal.

"The diagonal number in our case is 0.8025142 . . . a_n . . . .

"Here is what Cantor did: he went about changing this diagonal number, one digit at a time. Now there are, of course, many ways to change a

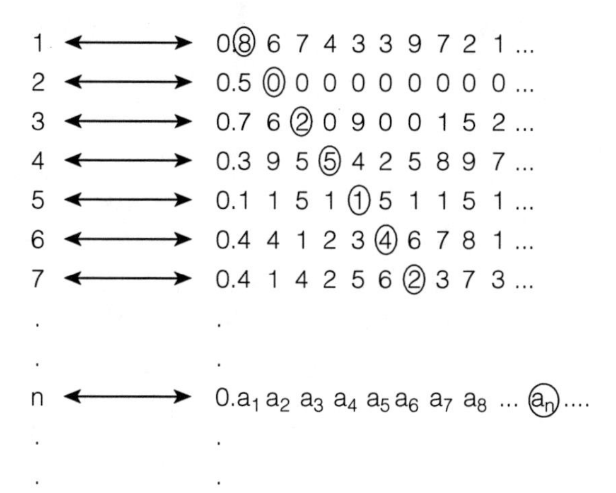

digit. I choose to subtract 1 from the diagonal digits with the convention that 1 taken from 0 is 1. So let us write out our changed diagonal."

The changed diagonal number in our case is 0.7114031 . . . ($a_n - 1$) . . .

"This changed diagonal number cannot appear matched to any integer. It can't be matched to 1 because we have changed the first digit from 8 to 7; it can't be matched to 2 because we have changed the second digit from 0 to 1; it can't be matched to 100 because we would have changed the 100th digit in the decimal expansion. No matter what integer 'n' you say this changed diagonal is matched to, you'd be wrong because we'd have changed the nth digit in the decimal expansion that is matched up to n. So this changed diagonal number cannot have any integer matched to it. And this is completely independent of the exact one-to-one correspondence. In my case I matched 1 with 0.8674339721. . . , but there is nothing in the argument that changes if you had matched 1 with some other real number. The same argument will work for any correspondence you could come up with!"

Nico's voice had risen several octaves as he built the diagonal argument. "Do you see what happened here?" he asked with rising excitement. "We started by saying that all the real numbers between 0 and 1 were one-to-one matched with the integers. Then we constructed a real number that could not possibly be matched to any integer, which contradicts our original premise of being able to construct such a mapping in the first place. We're forced to conclude that such a correspondence is impossible, and consequently there are more real numbers than integers."

There was an intense silence in the room. Every student in the classroom was looking hard at Nico's picture. I found myself going over the argument trying to grasp it in its entirety, but it kept slipping away. I couldn't quite see it fully.

Peter gave voice to the question that was bugging me. "Why not simply place the changed diagonal number opposite the integer 1 and move each number in the list down by one position?"

Nico laughed. He had the feel of a kid playing with a puppy in his backyard. "Peter, you could try that, but I'll just wait 'til you're done and construct a new changed diagonal number working with the new list. My new changed diagonal number wouldn't be matched to 1 or any other number because we would have once again constructed it not to match any integer!"

Slowly, like ink spreading through a jar of water, I began to become aware of the inevitability of Cantor's process. The method was indeed like Claire's "no greatest prime argument" or Pythagoras' demonstrations showing that $\sqrt{2}$ is not a fraction. You started by assuming the opposite of what you wanted to show and derived a contradiction. But I thought the specifics of Cantor's argument were particularly elegant. No matter what you tried you couldn't possibly match the whole numbers to the reals. So you *had* to believe that there were more reals than integers. And since there were as many integers as rational numbers you could also safely conclude that there are more reals than rationals. Infinity, just as Nico had promised, was yielding its mysteries.

But Claire had a wrench I had not considered. "Why wouldn't the same argument work for rational numbers?" she asked. "I mean, you could lay out the decimal expansion of rationals, match them up with the whole numbers, and then repeat the diagonal argument. Wouldn't you end up proving that there are more rational numbers than integers?"

"Claire, let's think about that. Suppose each integer is matched to a rational number and you then try to apply the diagonal argument. You construct a changed diagonal by altering the first digit of the decimal expansion of the first rational, the second digit of the second rational, and so on. Notice that there is no way to ensure that the changed diagonal number you construct will be a rational number at all. For the changed diagonal to be a rational number, it would have to either terminate or repeat. But this requirement forces our hand and we may not be able to change a particular digit on the diagonal when we need to do so in order to retain the terminating or repeating structure we had been developing to that point."

Claire thought about that for a minute. Slowly she nodded. And then she smiled—a slow smile that lingered for a while. I remember thinking about that smile later that night. Maybe she smiled because she ruefully realized that she should have figured the answer out for herself. Or maybe she smiled because she noticed that every time she said something I would turn around and look at her in anticipation and admiration.

• • •

April 24, 1919

Vijay Sahni vs. People of New Jersey

Official Court Document

Judge Taylor: Mr. Sahni, good morning.

Vijay Sahni: Good morning, Judge. If I am not mistaken, you have a lot of geometric figures drawn in your notebook.

JT: Yes, indeed. The Pythagorean demonstration you did for me awakened memories that have been dormant for over four decades. But I have a great many things to talk to you about today, and while geometry is one of the areas we will touch upon, I am most interested in understanding the connection between your mathematical thinking and what you said in the town square. *[Note from Court Reporter: Prisoner attempted to interrupt Judge's statement.]* No, please do not respond. I have already planned a structure for our conversation and I intend to stick to it. While I found our previous conversation interesting, even illuminating, I think it was poorly structured, and I would not like to repeat that error today. So I encourage you to listen attentively. I assure you there will be a full opportunity for you to make any points you would like to make to support your arguments.

[Note from Court Reporter: Prisoner requested a sheet of paper and a pencil to take notes on Judge Taylor's statement. Request granted by Judge Taylor.]

VS: Very well, please begin.

JT: As I indicated, after I left here last week, I reviewed the construction you drew out for me. I remembered studying the theorem in school, and although my recollection was blurry, I did not remember the demonstration as being as uncomplicated and simple as the one you showed. I actually pulled out my textbook from 1874 and I was proved right. The demonstration was quite complicated and it did not seem to shed much light on why the result was true. In fact, the demonstration was completely ignored by our teacher, a Mr. Davis from upstate, which as it turns out was his habit. My class notes, still legible after all these years, were filled with result after result with nary a line about why any was true in the first place. At that time I saw geometry no differently from Latin, where rigid rules control what is permissible and what is not. But I now began to glimpse that you see geometry as a set of understandable and demonstrable truths rather than a set of postulated rules.

VS: You are completely correct, Judge. Sorry, please continue.

JT: At any rate, my interest piqued, I went to the library and saw several results demonstrated in the manner you began to show me. There was theorem after theorem, each with a reason and justification, a method and a purpose, and I sensed the cold and austere beauty of the demonstrations, with not a wasted step or movement, a stern perfection unparalleled by any science or art that I have seen. As night fell I asked the custodian for the key, promising that I would lock up when I left that night. But when he walked in the next morning he found me where he had left me.

VS: What did you read that night?

JT: I read all the geometry I could lay my hands on. Primarily I was interested in understanding the method of proof because that really seems to be the source of your argument against faith. But along the way I confess I did become fascinated by the demonstrations, and read them as much to satisfy my own curiosity as to serve the purposes of this case. I started with the Pythagorean theorem and saw several proofs, but not the one you showed me. One of the proofs, it may interest you to know, is credited to the American President James Garfield, who in 1876 measured the area of a trapezoid using two different methods.

VS: He was really a president?

JT: Indeed he was. And he was a Christian; apparently mathematics did not create a dichotomy for him. Be that as it may, I was telling you about my night at the library. I read many theorems and their demonstrations and I could glimpse that mathematical reasoning is different from the ways of faith, but I couldn't quite put my finger on what precisely made it different. Finally I decided to try a theorem for myself, believing that the process of creating the demonstration would shed light on the matter.

VS: What theorem did you pick?

JT: A theorem that intrigued me the minute I saw it. It was in one of the books I was reading. When I read the statement I immediately was awed by it. It seemed to be an amazing fact and I decided that this was a result I would try to prove. The theorem states that "the angle subtended on a semicircle is always a right triangle."

VS: Sorry, subtended? What does the word mean?

JT: The theorem is saying that if you draw a semicircle and take the diameter as two vertices of a triangle and any point on the semicircle as the third vertex, then the angle made on the semicircle will be a right angle. Here, let me draw a picture.

[Note from Court Reporter: Judge Taylor's drawings are reproduced below.]

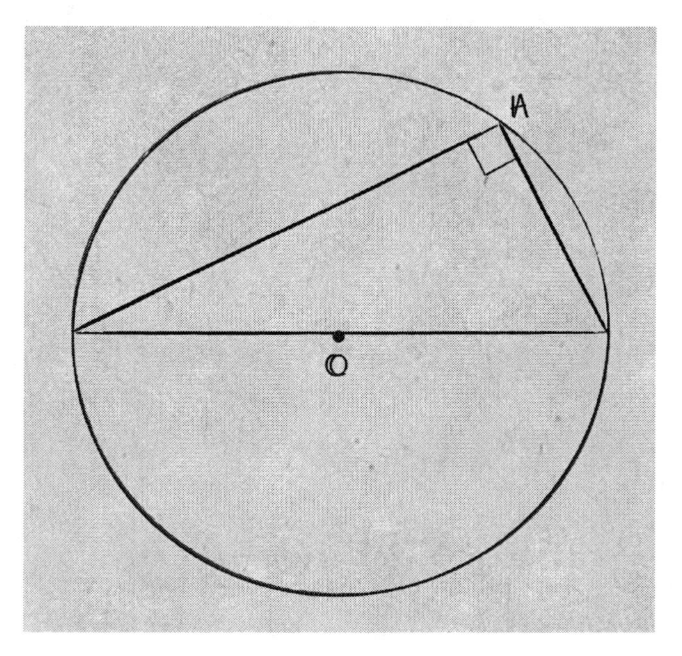

JT: So here is a circle with a center at O. The theorem is saying that the angle at point A is a right angle, no matter where you draw A—as long as it is not drawn at the two points at which the diameter meets the circle.

VS: Yes, of course! I know this result. Were you able to prove it?

JT: Prove it? At first I didn't even understand it. I thought that the author of the book intended something unique about Point A and that if we chose any other point on the circle we would not get a right angle! I must have spent quite a lot of time trying to figure out what was unique about Point A. Suddenly I realized that the author was saying that the property held regardless of where "A" was. It was this drawing that cleared it up.

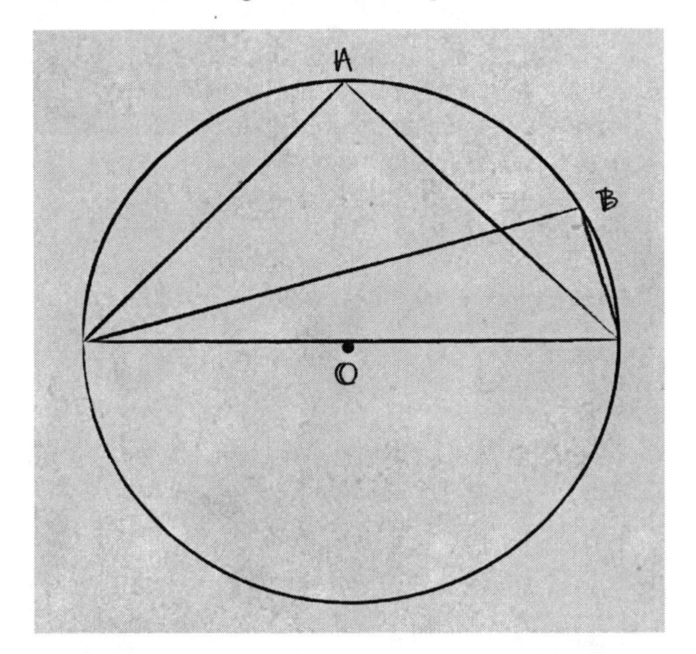

JT: It was with considerable surprise that I realized that the theorem was saying that whether the triangle touches the circle at "A" or at "B," it always makes a right angle. Once I realized this, I was even more in awe of the theorem. I kept telling myself that there had to be a reason. Why should a triangle on a circle always have a right angle? It seemed to me to be a great mystery, one that I was determined to resolve.

VS: Clearly you were seized by the wonder of thinking, Judge! Please tell me, how did you approach the problem?

JT: For many hours, I didn't quite know how to approach the problem. I tried reading the other theorems proved in the textbook for inspiration, but nothing seemed to apply. The one theorem I thought I could use said that the base angles of an isosceles triangle are equal. An isosceles triangle has two sides that are equal.

VS: Yes, I know. What interests me is how you had this insight about applying the isosceles triangle theorem.

JT: I don't know. After I completed the solution, I wondered the same thing. Perhaps it was because I tried everything, and this was the only result I found applicable. At any rate, here was the picture I drew, and may I say that it filled me with a good deal of pleasure.

[Note from Court Reporter: Drawing reproduced from Judge Taylor's notes.]

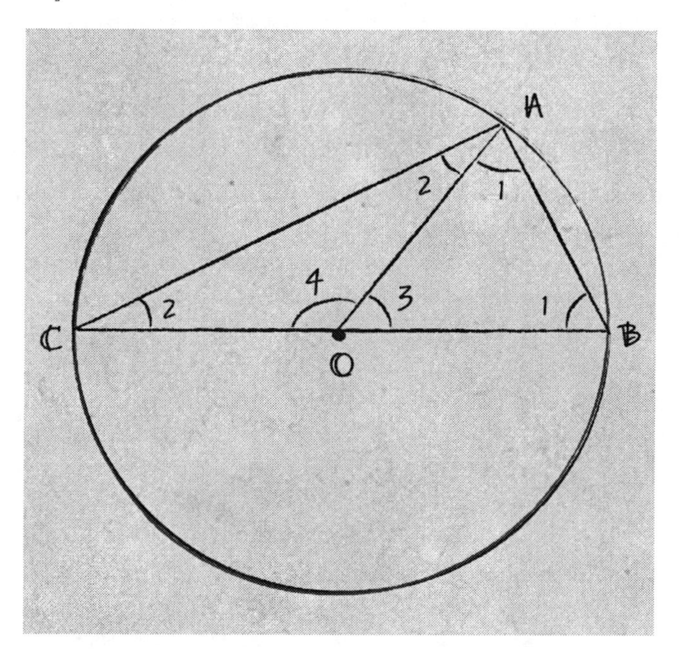

JT: As you can see I drew a line from the center "O" to the point "A" on the circle. This line, which I'll call OA, is the radius and it splits the angle at A into Angle 1 and Angle 2. We are trying to show that Angle 1 + Angle 2 = 90°.

VS: Precisely.

JT: I used the isosceles base angle theorem and the fact that OA = OC to show that the two base angles are equal. Since the angles are equal I named them both Angle 2. A similar pattern holds in the triangle OAB. Both OA and OB are radii of the circle and so their base angles are equal, and I named them Angle 1. Since the sum of the angles in a triangle is 180°:

Angle 1 + Angle 1 + Angle 3 = 180° and Angle 2 + Angle 2 + Angle 4 = 180°.

By adding both equations we can see that:

2(Angle 1 + Angle 2) + Angle 3 + Angle 4 = 180° + 180°.

But since Angle 3 and Angle 4 are the angles on a straight line, their sum must equal 180°. And so: 2(Angle 1 + Angle 2) = 180°, and dividing by 2 we get, Angle 1 + Angle 2 = 90° which is what we were trying to demonstrate!

VS: Judge, that is wonderful! You truly have a fine brain—a mathematician's brain. Congratulations!

JT: Thank you. But my investigations were to yield more important riches. Just before dawn, perhaps spurred into an altered state due to lack of sleep, I had a critical insight. It came to me that what was unique about mathematical thinking is that every new conclusion follows from a previous simpler fact. For example, I could show that the angle on the circle is a right triangle by using two "simple facts": one, that the base angles of an isosceles triangle are equal, and two, that the sum of the angles in a triangle is 180°. Just by using these facts, the angle on the circle had to be a right angle. The proof of the Pythagorean theorem you showed me was the same way. It used "simple facts" about computing the areas of squares, rectangles, and triangles to prove the ultimate result. Indeed, every theorem I've seen in the last few days has relied upon previous "simple facts." Do you agree with this conclusion of mine?

VS: Yes.

JT: I thought as much. I'm enjoying this, Sahni! As soon as I gripped this truth about mathematical results relying on previous "simple facts," I asked myself how this method was different from the ways of faith. And it occurred to me that the difference was purely stylistic. Faith can be based on the very same deductive methods that mathematics comprises. For example, I can prove the existence of God using purely deductive methods starting only with "simple facts" of my own.

VS: Judge, if you can do that, then I will willingly acknowledge my error and plead guilty to this charge before me.

JT: Your eyes tell me that you are certain that I will not be able to do that. Anyway, let me sketch out the idea that occurred to me that morning at the library: I started with a "simple fact" that everything in the universe must be created by something; it cannot come into being out of nothing. I have heard evolutionists say that man arose from animals, and I know that they will argue that animals came from fish and that the fish came from water and the water came from the earth, and perhaps that the earth came from the sun. I have read all these theories. But the fact remains that there must have been a "first something" that did not arise from something before it. Now by my "simple fact" about everything in the universe being created, this "first something" had to have been created as well. But since it was the "first something" it must have been created by a being that was always there, a being that could create something from nothing, and a being that lived before time started. Humans have just given that being a name: we call him God. So you see, Mr. Sahni, the fact that anything exists at all is proof that a creator exists. Do you see what I'm getting at?

VS: I understand your argument quite well. In fact I have heard it before. The problem is that I do not believe your underlying "simple fact" that everything in the universe must have been created by something.

JT: How can you disagree with that? It is crystal clear.

VS: Everywhere I look, life is full of cycles. Evaporating oceans create the rain and in turn the falling rain creates the oceans. Perhaps the "first thing" you speak of is part of some similar unending cycle.

JT: But someone had to create the cycle! I'm telling you that God is the cause of the cycle. He is the first cause of everything, and all things must have a first cause.

VS: Why, Judge? I could equally well claim that the "first thing" always existed. Your underlying assumption that everything must have a cause of creation applies to the world of sense experience. In this realm I will agree with you that your "simple fact" of causality is self-evident. But you have extended the "simple fact" beyond the world of sense experience to something that is supposed to transcend it. You have applied an experiential notion beyond all possible experience, as well as beyond the limits for which there are any guarantees that our sensory perceptions are reliable. So, yes, Judge, I disbelieve your underlying "simple fact," and hence I disbelieve your conclusion that God exists.

JT: You know, I could equally well say that I disbelieve the simple fact that an isosceles triangle has equal base angles.

VS: No, no Judge, you are wrong! The base angle theorem is provable. It logically follows from other simple facts.

JT: But each result cannot follow forever from other simple facts. At some point something must be taken as a given.

VS: You are correct. In mathematics, these givens are called axioms. An axiom is a self-evident beginning principle—something that is clearly true from which you can deduce new truths. From these axioms simple facts are derived, and from these simple facts more complicated results follow. For example, the fact that the sum of the angles of a triangle is 180° is a simple fact that logically follows from simpler axioms. You used this simple fact to derive the more complicated result about the angle on the circle. Euclid, a Greek geometer, and in my opinion the first modern mathematician, derived all of geometry, including the Pythagorean theorem, from a small set of obviously true axioms.

JT: Obvious in whose eyes?

VS: In the eyes of any intelligent being who evaluates them.

JT: What if I disbelieve them just the way you disbelieved my axiom about everything having been created from something.?

VS: You will not honestly be able to disbelieve Euclid's axioms. You are an honest man and I have complete confidence that you will not allow our philosophical differences to cloud your judgment.

JT: But sir, you are letting your bitterness against faith keep you from seeing my axiom that everything around us must have a first cause.

VS: Not at all, Judge. Your "axiom" as you call it, fails the test of self-evidence, so I do not accept it as an axiom. I will show you Euclid's axioms and you will see that they are truly self-evident. Euclid started with the axioms as a foundation and built forward by reason, and reason alone. Therefore, no doubt was allowed to shade anything he claimed to be true. To doubt Euclid you have to question either the axioms themselves or the reliability of reason, neither of which allows any room for questioning.

I believe that every field of human knowledge must be axiomatized and the laws of that body of knowledge must be deductively derived from the beginning axioms of that field, be it physics or human behavior. And it is only by applying these methods that we can get to the truth. All else is vanity. It was Descartes in 1637 that first voiced his expectation that truths in all branches of knowledge will be acquired by mathematical methods. There is a beautiful section in his *Discourse on Method* which I've read so many times that I could recite it word for word. Here is what he wrote:

> The long chains of simple and easy reasoning by means of which mathematicians are accustomed to reach conclusions of their most difficult demonstrations led me to imagine that all things, to the knowledge of which man is competent, are mutually connected in the same way, and there is nothing so far removed from us as to be beyond our reach or so hidden that we cannot discover it, provided only we abstain from accepting the false from the true, and always preserve in our thoughts the order necessary for the deduction of one truth from another.

In the centuries after Descartes, his vision has proven to be correct. Mathematical methods have had success in all branches of learning from physics to philosophy, from music to morals, from the laws of wealth to the laws of human society. And all of these methods begin from axioms that we know to be true beyond any reproach.

JT: I would like to judge that for myself, Mr. Sahni. When you show these axioms to me I will judge if they rest on ground any firmer than my axiom, and I will continue to refer to it as such. I daresay that we have come to the very heart of the matter as we discuss the truth of these axioms. I would love to move this to completion right now, however some urgent business takes me to Boston this afternoon. We shall continue upon my return next week.

VS: Very well. In the interim, may I request a book from the library?

JT: You may. What is the name of the text?

VS: *The Elements*, by Euclid.

JT: Ah! I should have thought as much. I recently have read that Abraham Lincoln greatly admired *The Elements*. It is said that long after his associates had retired, Lincoln would stay up following the theorems of Euclid. Mr. Hanks here will ensure that you receive a copy of the book as soon as possible, and I shall look forward to hearing about Euclid's schema the next time we meet.

VS: Yes. Thank you.

JT: Oh and one last thing.

VS: Yes?

JT: It may interest you to know that Descartes was a devout Christian.

VS: I know that. But . . . *[interrupted]*

JT: I'm sorry but I must leave now. Good day, sir.

• • •

Bauji was saying that Euclid's geometry could be reduced to a simple set of axioms from which everything would follow. And he seemed to believe that this axiomatic model was somehow applicable to all aspects of life. For my part, I was not even clear on how exactly you could axiomatize geometry, to say nothing of axiomatizing the more ambiguous, philosophical questions that life confronts us with.

Unable to sleep I had stayed up all night reading the transcripts. I was trying to get a sense of how Bauji could possibly connect mathematical axioms to his notion of building certain truth. Moreover, how did the Bauji of these transcripts evolve into the man I knew? If axioms are so important, why had he never talked to me about them?

Things were no clearer in the morning. By the time Claire called asking if I wanted to go out on a run, I was more than ready for a distraction, and she was as good a distraction as I could possibly get. So once again, I found myself panting up the Stanford foothills (by the end of the semester I had gotten into the best shape of my life). Jogging was the only time when Claire was completely relaxed and open, and a daily reminder of my relative aerobic inadequacy was a small price to pay for the pleasure of her company.

When I told her about my brooding over what Bauji might have meant, she looked at me rather quizzically. "I know these questions are important to you Ravi, and you should probably run them by Nico. When I hear you and Adin talk about these issues my curiosity is aroused, but I don't feel the same urgent need to find the answers. To me the question of the foundations of mathematics or human thought is academic. I'm motivated to actually do mathematics. There is a whole edifice of knowledge standing there that I want to add to. Like building a new room in the building, a room that looks out onto something new. The foundations don't worry me; the very fact that the whole edifice exists seems to suggest that the foundations are rather solid."

She did, however reluctantly, accompany Adin and me when we went to meet Nico at his office. I did most of the talking, beginning with Bauji's visit to Morisette and ending with the questions about what axioms really meant and how they applied to mathematics and life in general.

Even as I spoke, Nico kept flipping through the transcript. "It's a fascinating story Ravi. I would love to know how it ends. And what is most amazing is that the two streams of mathematics that you find yourself immersed in—Euclid's geometry and Cantor's theory of the infinite—have fascinating and startling parallels. I can't tell you exactly what those parallels are without developing these subjects a bit more, but in due course, you will see that both these subjects take us to the farthest boundaries of thought; both make us think about what it really means to be certain about an idea."

"Now, this thing with the axioms . . ." Nico paused here, letting out a heartfelt sigh that seemed interminable. "It's a really delicate issue. The best way to think about axioms is that they keep us honest. You start with a few axioms and you logically derive new truths. If you state as an axiom that only clouds bring rain, and looking outside your window you observe that it is raining, you may conclude by the laws of logic and your axiom that there are clouds outside. But you may not conclude that there is a rainbow. You have no axioms about a rainbow and your conclusion wouldn't have a foundation. So mathematicians use axioms as "starting truths." They allow us to get started. After Euclid every branch of mathematics has used axioms."

"I have a question," said Adin. "How come the Cantor stuff we're doing in class didn't begin with any axioms?"

"Cantor's mathematics *does* have axioms. They were set forth after Cantor developed his theory of infinity."

"Wait a minute. Why did the axioms come after the theory? Shouldn't he have started with axioms?"

Nico laughed. "You'd think so, wouldn't you! But that is not how mathematics works. For many mathematicians axioms are not important. There is a pervasive belief that as long as a hypothesis produces results and there are plausible demonstrations to show it is true, then mathematics can and should proceed."

Claire, who had been doodling a pentagon into a circle looked up and nodded. "That's what I think. I know a proof is true when I see it and I don't need any axioms to be certain about it. For example, I believe that there is no largest prime—and the demonstration I've seen didn't need any axioms."

Nico nodded. "Most mathematicians think the same way, Claire. They feel a proof in their bones and don't feel the need for any axioms. But every so often they get trapped in a logical impossibility—a paradox. One of the main reasons mathematicians began to axiomatize mathematics was to avoid paradoxes from occurring."

Claire nodded, but she seemed to do so without enthusiasm. Adin, however, was at the opposite end of the spectrum. Given his quest for certainty these axioms could well be the key he had been looking for. To do mathematics by ignoring axioms was not something he would ever approve of. And so, not surprisingly, he hung on to every word that Nico was saying.

"So axioms are created to avoid us picking up false assumptions along the way?" he asked.

Nico nodded. "I'd say that's fair. It's easy to pick up a false assumption and not even know it until you get hit by a catastrophic paradox."

I kept thinking back to Zeno's argument, which at least in part may have been based on the untrue axiom that infinite sums must diverge to infinity. I thought that the paradox occurred because Zeno's axiom was false.

Nico didn't quite agree. "Ravi, Zeno didn't have any axioms. He just naively made an argument that was unsupported by axioms. When we looked at his argument we essentially stripped it bare and looked at his hidden assumptions, which is where we found the problem. This is why axioms are important. They force us to be explicit about our assumptions, and it allows people to judge if they are evidently true or not. It is the only way to guarantee certainty."

I started to see how this whole axiom thing worked. Axioms had to be simple self-evident truths. If you used them properly you would end up with truths you could be certain about.

But Claire was not buying it. I saw her tie her hair into a pony tail (a sure sign of her getting serious) while shaking her head in disagreement. "Zeno was wrong because he didn't do all of his mathematics correctly. Had he been clear and explicit about everything there would have been no paradox."

Nico shook his head in turn. "It's not that simple, Claire. Let me see if I can come up with some appropriate examples," Nico said looking toward his window for inspiration. This was different from his lectures in class. There he had a structure laid out, a structure he had already gone over in his head, innovating only in response to questions. Now he was working on a theme, creating something anew at every moment.

"Ah yes, let me take up what I like to call the postmodern paradox. It fits in well with what we will take up in the next few classes. Imagine an old-fashioned scholar of English literature weighed down by postmodern tendencies, yet with a need to make a living in this new world. So he embarks on a project that makes sense to his more fashionable colleagues, yet seeks to repudiate some of their work. Heartily sick of modern novels where a character on page 60 is reading through the very book he or she figures in, he decides to compile a book that lists all those books that do not refer to themselves. And then, in the course of his work, he runs into a paradox: Will his book include itself or not?"

The nature of the paradox came to me all at once. If his book did not refer to itself, then it was a part of the list of all books that did not refer to themselves and it therefore belonged in the list our author was compiling. But by being in that list the book was referring to itself, which was opposed to the premise we started with. If on the other hand we began with the premise that the book did refer to itself, then it was already in the author's list of books that do not refer to themselves. It was a maddening dilemma.

"But before you start thinking all this is just a matter of word puzzles and logical games, it isn't so," said Nico. "This is somewhat of a contrived example but it comes from a real problem in Cantor's mathematics. Without axioms Cantor's theory would have fallen prey to a similar contradiction. And in a couple weeks I'll show you this contradiction in class."

"So what exactly is wrong with a contradiction?" asked Claire.

"Fair question," said Nico. "A contradiction raises the frightening prospect that all of mathematics is shaky and that no proof is reliable. It completely washes away any possibility of certainty about anything. Accepting a contradiction allows us to prove whatever we want. Say there is a statement S that is both true and not true, for this is what a contradiction is. Here is how you can show anything from a contradiction:

> If S is true, and Q is any other statement, then "S or Q" is clearly true. Since S or Q is true and S is not true, then Q is true, no matter what Q is.

So if a contradiction exists anywhere in the universe, then everything is provable! There's a funny story about this. Once someone asked the logician Bertrand Russell whether, given the contradiction that 1 = 2, he could prove that he was the Pope. The quick wit that he was, Russell is said to have instantly responded: The Pope and I are two, and since two is equal to one, the Pope and I are one. So you see I am indeed the Pope."

Claire laughed, a happy sound.

"I have a question," said Adin. "Is the purpose of a set of axioms to help us avoid getting to contradictions?"

"One would hope so. An axiom set that leads to a contradiction is worthless. In general, axioms help us avoid falsehoods. A contradiction is a symptom of a prior falsehood."

Adin nodded vigorously. He seemed in danger of exploding unless he could give the right words to his thought. "I think I see the big picture. Contradictions are not permissible in mathematics but they occur in life all the time. People routinely live contradictions and they can do this because they have no axioms that form the basis of their thoughts. As you said, Nico, the job of the axioms is to keep us away from contradictions. That is why you need axioms in philosophy just like you need axioms in mathematics."

"The people of Morisette would say that the Bible has all the axioms you need in life," I told Adin.

But he had already thought of that. "As your grandfather points out, axioms must be self-evident. The Bible's axioms are far from being self-evident. In fact, some of them are contradictory. We need a philosophical system that springs from noncontradictory, self-evident axioms. That alone will take us to certainty in human ideas."

But Nico shook his head. "I think it's more complicated than that."

CHAPTER 1 2 3 4 5 6 7 8

EUCLID — JOURNAL ENTRY, DATE UNKNOWN

Visiting with Tantalos yesterday an amazing thing occurred. He was in his customary frenzy—demonstrating proposition after proposition with great flair and alacrity. The usual circle of geometry enthusiasts had gathered around him, spellbound by his genius and oratory. As is becoming more and more typical, they completely ignored me. They see me as a relic of the past whose mathematical talents are but a shadow of Tantalos'. So I just sat quietly for most of the evening, left alone to listen to him and absorb the torrent of knowledge that gushed from him.

And then, as evening fell, it happened. Tantalos demonstrated a proposition by relying on a proposition that itself relied on the original proposition for its demonstration! A snake eating its own tail! In essence, he demonstrated Proposition A by assuming the truth of Proposition B, but a few hours ago he had demonstrated Proposition B by assuming the truth of Proposition A. His argument was more complex and the context had many interfering details, but in essence that is what he had spun—an argument as circular as a chariot's wheel.

I pointed this out, quietly but firmly. At first the crowd dismissed my challenge; some looked at me pityingly, as if I knew not of what I spoke. How could old Euclid be questioning the best geometer in Alexandria? But I persisted, and slowly the tide turned. In the court of geometry even mighty reputations must yield to the power of reason.

When Tantalos himself acknowledged the truth of what I was saying, an awkward silence gripped the group. It was unclear if either Proposition A or Proposition B is even true. Such doubt is permissible in the musings of the medical men, who change their treatments without notice or reason, but in geometry, doubt is fatal.

Tantalos tried to rectify the flaw, but he could not. Indeed, he had no method to proceed. Even if he could show Proposition A without assuming B, what would guarantee that there would not be some other undetected circularity in his thinking?

I brooded over this. Tantalos may be a faster thinker, but I believe that my brooding has its own rewards. This morning as I write these words, I have a strong sense that Tantalos' mistake has opened the door for my greatest work.

I realize that the only way to avoid circularity is to start by believing something. And this is what I will do. I will find some postulates that would unassailably be true, true beyond any possibility of doubt, and from these postulates I will logically and sequentially derive all the propositions of geometry. The postulates will give rise to simple propositions, and the simple propositions will demonstrate ever more complex results, until all of geometry, including all of the many theorems of Thales, Pythagoras, Eudoxus, and Theaetetus, will follow from the handful of postulates I will begin with.

Of course the most delicate work would be to pick the right starting postulates. They must be simple enough so that they are not demonstrable from even simpler postulates, apparent enough so that no one could doubt their certain truth for even a heartbeat, yet rich enough that they encompass all the results of geometry.

I will carefully write these postulates and demonstrations and be stubborn and uncompromising about rigor. I know Pythagoras thought about this approach, but for whatever reason he never executed it; but I will. This will be my greatest work. History will probably remember and revere Tantalos' brilliance, but perhaps it is my destiny to contribute certainty to human knowledge.

• • •

April 29, 1919

Vijay Sahni vs. People of New Jersey

Official Court Document

Judge Taylor: Good morning Mr. Sahni. I see that the library did indeed send you a copy of Euclid's *Elements*.

VIJAY SAHNI: Yes, they did. It was the first time I have read through Euclid's theorems the way he actually wrote them, and I must say that it was a wonderful experience. I have been eagerly awaiting your arrival today, for I fear that if I don't discuss the beauty and wonder of these results with you, I will burst!

JT: I appreciate your enthusiasm, Mr. Sahni, but let me caution you on our purpose: We are not here to discuss the beauty of Euclid's results. Rather my expectation is that you will show me a system that begins with certain self-evident axioms and, using only the laws of logic, deduces mathematical truths about the world around us. Therefore, I am expecting you to start with these axioms and I am expecting to see the use of logic to build the chains of reasoning that will lead to results that we can, borrowing your own words, "be certain of." And most importantly I would like to understand why you hold Euclid's axioms to be truer than the one I spoke of in our previous meeting, namely that everything in the universe must be created by something.

VS: Judge, I agree with your admonishment and understand your purpose. I will indeed show you a formal system, where nothing is left to intuition and every deduction rests upon previous results or absolutely certain axioms. I understand and respect the seriousness of our purpose. However, I suggest that you not close yourself off to the beauty and elegance of these results. For there would be little reason to build formal systems were they not beautiful—or useful.

JT: Very well. Please begin.

VS: Before we delve into the actual mathematics, Judge, I want to give you some context for Euclid's mathematics and why it literally changed the course of human history. I do this not merely because it is interesting, but because in a sense Euclid is the source of the philosophy that I have spoken to you about these last few weeks. Do I have your permission to proceed?

JT: Please do. I myself am curious about Euclid, the man. His name kept coming up that night in the library.

VS: To the best of my knowledge, Judge, very little is known about Euclid himself. We know he came to work at the Alexandria library around 300 BCE. We also know that he founded a school of mathematics and was quite familiar with the mathematical results of his predecessors. Not much more is known about his life. We know a few things he is supposed to have said. One story says that a student who had just learned his first theorem asked Euclid what he could gain by studying such things. Euclid is said to have asked his slave to give the student three obols, "since he must make gain out of what he learns." In another story Ptolemy is said to have asked Euclid if there was a quick way to learn geometry, to which Euclid is said to have replied, "there is no royal road to geometry." I do not even know if he actually said any of these things, Judge. For all I know, these tales are made up by historians with a flare for the dramatic.

JT: Why do you doubt everything? Why would a historian make up such tales? I think Euclid probably did make these statements. At any rate I reckon it's just your nature to doubt. Please continue.

VS: From his presentation in *Elements* it is clear that Euclid was a thoughtful, patient man, with a wonderful eye for detail. He was, of course, a very good mathematician, but he was probably an even better teacher. He wrote *Elements* so that people could study mathematics methodically. And such was his passion that he recorded almost all of the basic mathematics known at his time. He must have had a lot of energy, for *Elements* is a collection of 13 books that contain no fewer than 465 separate propositions from plane and solid geometry and number theory. Euclid's genius was not that he created every single one of these 465 propositions—indeed it is known that he borrowed heavily from the works of his predecessors. Instead, Euclid's genius was that he understood and illustrated the concept of proof. He founded, or at the very least propagated, the notion of mathematical rigor. He was passionate about certainty. He begins each book within *Elements* with a set of definitions and axioms. He then constructs a first proposition based exclusively on the definitions and axioms. Succeeding propositions build on previous results as well as the definitions and the axioms. And what emerge are beautiful, certain facts about the world around us. To this day mathematicians follow the same structure that Euclid laid out over two thousand years ago!

In a real sense, *Elements* has had an impact comparable to your Bible. When it was written, it accelerated the pace of Greek mathematics. It was translated into Arabic and it had a real impact on that culture. Translated into Italian, the *Elements* was one of the shaping influences of the Renaissance period. I know that Isaac Newton was an admirer of the book, and you indicated Abraham Lincoln himself was an ardent student of the *Elements*.

JT: Was it ever translated into Hindoo?

VS: Hindoo is the name of a religion. Hindi is the language many Indians speak. To my knowledge *Elements* was never translated into Hindi and I daresay we may have been a greater country had we had early access to such a translation. I first read a summary of the book written in English. My father got me a copy on my birthday after I made it clear that I wanted nothing else. It presented the theorems without bothering to lay out the postulates or the definitions.

JT: I see. Was this before or after you saw the snake-girl you told me about?

VS: Shortly after, Judge.

JT: I thought as much. But we shall set that aside. Please proceed with your presentation of Euclid's works.

VS: Very well. In Book I, Euclid starts with a set of 23 definitions, 5 postulates, and 5 common notions. These were his givens and from these givens alone he derived 47 increasingly complex propositions culminating in the demonstration of the Pythagorean theorem and its converse.

JT: The same Pythagorean theorem we demonstrated a few weeks ago?

VS: The very same, although Euclid's method of proof is different from the one we discussed.

JT: But we did not need any axioms, and it was a perfectly satisfactory proof.

VS: Judge, it appeared satisfactory because along the way we assumed many details without making our assumptions explicit. Let me give you an example. Do you remember the drawing?

JT: Vividly. As a matter of fact I could draw it from memory.

VS: Please do.

[Note from Court Reporter: Judge Taylor's rough drawing is reproduced below.]

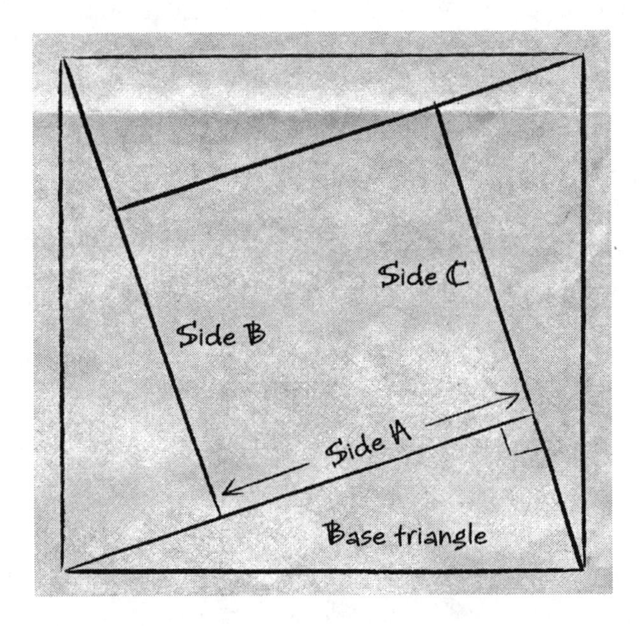

VS: Exactly correct. Now remember that our demonstration relied on the fact that the central figure is in fact a square.

JT: Yes. We had separated this figure into four triangles and the central square.

VS: Right. But let me ask you: How do you know that the central figure is, in fact, a square?

JT: It has to be a square, just look at it!

VS: It looks like a square but that does not mean it is a square. Drawings can be inexact. The eye brings illusions that seduce the mind into believing what is not true. I don't know if you are aware of the fact that the Greeks deliberately built the pillars of the Parthenon to converge very slightly towards each other—because that is the only way they could provide the illusion of parallel lines. To make railroad lines look parallel, artists converge . . . *[interrupted]*

JT: Yes, yes, I know. But look at the central figure. It has equal sides at right angles to each other.

VS: Judge, you are not seeing my point. Rereading the *Elements has* convinced me that looking at the figure can never be enough. Looking cannot lead to exact truth. Pictures can only aid in understanding and cannot by themselves lead to proof or truth. Only deduction and reason can lead to truth. Can you give me a reason why the central figure is a square other than the fact it looks like one?

JT: I see your point, but I think I can provide a reason. I am going to construct the drawing one step at a time.

[Note from Court Reporter: Judge Taylor's drawing is reproduced below.]

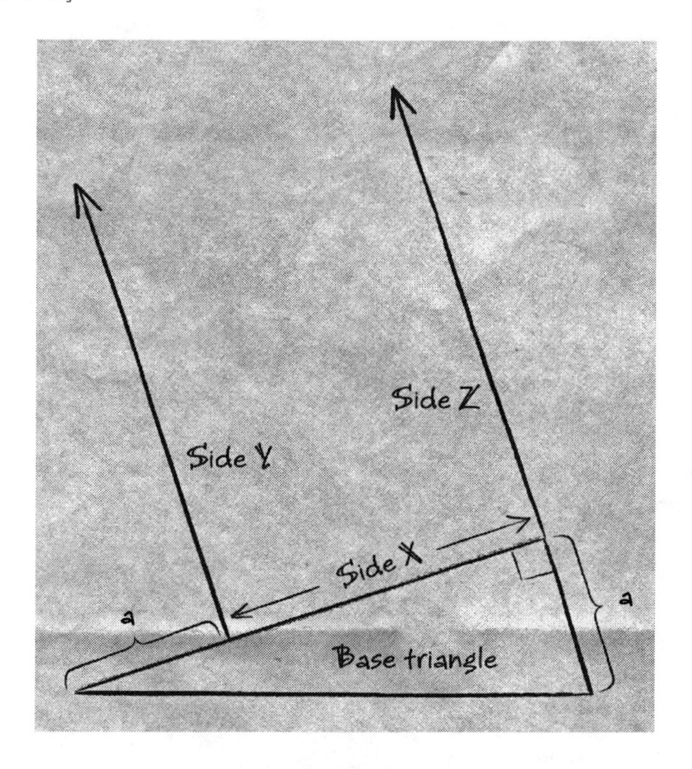

JT: This looks more complicated than it is, but it shows how I would go about doing the full drawing. All I've done is that I've started with the base triangle and I've extended the short side indefinitely. I've called the extension Z. And then I have drawn a line Y at a point that is exactly "a" distance away from the left tip of the triangle. Oh, and I have drawn Y to be perpendicular . . . *[interrupted]*

VS: Judge, I'll have to stop you here. I know you don't realize this, but you have already made three assertions that must be proven, or else assumed to be axioms.

JT: Three assertions? What do you mean?

VS: First you have assumed that a straight line may be extended—that's how you drew your line Z.

JT: Of course you can extend a line—just use a straight edge!

VS: I agree with you. All I am pointing out is that in effect you have introduced an axiom. And it is fine to introduce axioms as long as one explicitly recognizes and records them. Otherwise one may make dozens, even hundreds of assumptions, never knowing what one has assumed, and what one has proven.

JT: I'll concede that. So what else—you mentioned there were three things?

VS: Yes. Secondly, you said you could mark off the point that is "a" distance away from the vertex of the triangle, perhaps by measuring and transferring distances.

JT: Yes, you can easily do that with a compass.

VS: Yes you can—certainly with a compass you can transfer distances. And it is quite OK to use this as a second axiom, but again let us recognize it as such. Thirdly, you had started to draw a line perpendicular . . . *[interrupted]*

JT: Yes, you can do that with a protractor. So, I reckon that is my third axiom.

VS: It could be. I will point out that Euclid never needed a protractor. He built all of his geometry without having an angle-measuring device.

JT: Why would he do that?

VS: Because frugality and simplicity were a virtue for him. His instincts made him seek the fewest number of simple postulates that he could build geometry on. He built his entire geometry on five postulates. You've already got three axioms in your theorem, and you're just getting started.

JT: Five axioms? That's all?

VS: Yes.

JT: How did he decide if something should be an axiom?

VS: He chose them with great care and foresight. He wanted to assume only the unassailable—that which is absolutely clear to our minds to be true. And he wanted to assume the minimum possible, for he did not want to call something an axiom when it could easily be proved from simpler axioms.

JT: Well, what were they, and how did he build all of geometry on five axioms?

VS: This is what I will describe to you, and in doing so I will show you an entire system of knowledge that starts with five undeniable truths as axioms and builds theorem after theorem, each one of them absolutely and completely certain. These theorems will be pure truth, unfettered by lapses due to emotion or longing.

JT: Longing?

VS: When people want to believe something strongly enough, they accept it as truth, even if there is no justification for doing so.

JT: I see. Doubtless you believe my faith to be such a phenomenon. *[Attempted interruption from prisoner]* No, no, there is no reason to get into that. Let us finish with Euclid first. We have completed the preliminaries and I would now like to completely understand this type of reasoning that in your mind renders the ways of faith to be misguided.

VS: Very well. We shall begin. Since this is quite possibly our last mathematical discussion and my last chance to articulate my world view to you, I ask that you listen with an open mind.

JT: I always do, Mr. Sahni, and on this occasion, too, you have my word.

VS: Excellent. Well let us begin then. Euclid starts with a set of 23 definitions. Definition 1 reads, "A point is that which has no part."

JT: Without geometry there is no point.

VS: What?

JT: An attempt at a joke. Never mind, please proceed. *[Note from Court Reporter: Judge Taylor waved off the prisoner's confusion and asked him to continue.]*

VS: Okay. Definition 2 states that "A line is a breadth-less length," and Definition 3 notes that the "ends of lines are points." Notice that these definitions are somewhat imprecise.

JT: Yes, they are. The definitions seem vague.

VS: Every system must start somewhere and not every term can be defined in terms of other elements in that system. A system where everything is defined must be circular. Euclid avoids circularity at the cost of vagueness.

JT: Yes, you need to start somewhere. I assume the next definitions are firmer?

VS: They are. Take Definition 10, for instance: "When a straight line standing on a straight line makes the adjacent angles equal to one another, each of the equal angles is right, and the straight line standing on the other is called a perpendicular to that on which it stands." It is interesting that Euclid does not define a right angle to be 90°. As far as I can tell, all of *Elements* refrains from mentioning any angle other than the right angle.

As you see *[Note from Court Reporter: Prisoner points to list of definitions in* The Elements*]* in these subsequent definitions, Euclid goes on to define triangles, circles along with their centers and diameters, quadrilaterals, and he later specifies the definitions of equilateral and isosceles triangles.

JT: Yes, I recollect these terms quite well. An equilateral is a triangle all of whose sides are equal, and an isosceles triangle has two sides equal to each other.

VS: Correct. Moving on, Euclid defines several other figures including circles, squares, rhomboids and trapezia. Let us take a look at Definition 15, for example: "A circle is a plane figure contained by one line such that all the straight lines falling upon it from one point among those lying within the figure equal one another." What Euclid is saying here is that a center point and a sweeping radius define a circle.

JT: Now that I think about it, that is indeed what a circle is. But it seems a long way to go to define something that is already familiar to everybody.

VS: Certainty demands precision, and precision demands explicitness. This is true in every walk of life.

JT: Keep going.

VS: Most important, however, is Euclid's last definition, Definition 23, where he defines parallel lines: "Parallel straight lines are straight lines which, being in the same plane and being produced indefinitely in both directions, do not meet one another in either direction."

JT: Yes, that seems logical.

VS: Indeed. Given these 23 definitions, Euclid moved on to his axioms. He referred to his axioms as postulates. Euclid could have chosen any number of postulates, but in his choosing the way he did he truly showed his genius. His postulates certainly meet the criteria that we had set forth for the axioms: they are certainly self-evident. But additionally they are frugal and avoid overlap. Let me write out the first three postulates:

[Note from Court Reporter: Prisoner's writings reproduced below]

Postulate 1: [It is possible] to draw a straight line from any point to any point.

Postulate 2: [It is possible] to produce a finite straight line continuously in a straight line.

Postulate 3: [It is possible] to describe a circle with any center and radius.

That is almost exactly what Euclid wrote over two millennia ago! I put the phrase "It is possible" in parenthesis because literally translated, Euclid's postulates do not include that clause. So the first postulate reads; "To draw a straight line from any point to any point."

JT: I see. You realize that these postulates encompass everything you can do with a compass and a straight edge.

VS: Almost everything, Judge, but not everything. With a real compass you can transfer distances by leaving the compass open at a particular length and draw a line segment of that chosen length elsewhere on the plane, which I recall is an axiom you wanted in your earlier demonstration. But Euclid does not allow constructions of this type. It is as if his compass collapses as soon as you lift it off the page.

JT: Are you saying that you cannot transfer lengths in Euclid's system?

VS: Judge, as has become your habit, you have once again asked the next logical question. You can indeed transfer lengths in Euclid's system merely by using a collapsible compass, and Euclid demonstrates a mechanism for doing so in the third proposition of Book I. Now, it is true that he could have merely assumed a fixed compass as a postulate and he would have been well within his rights to do so. But such were his instincts for organization and economy that he refrained from doing so! He did not want to assume a fact as a postulate when he could prove it using other postulates. This is a small instance of his genius.

JT: So, he assumed the least that he could have assumed in order to achieve his purposes. But what were his purposes? Did he know what they were before he began writing his books?

VS: It is impossible to say, Judge. My impression is that he wanted to rigorously organize the various results that were known to his predecessors.

JT: So he made no original contributions?

VS: Oh no, not true at all! Even if every single proposition in *Elements* was known before him, it was Euclid who discovered the imperative for rigor. This is his greatest contribution, and I would say that it is the greatest contribution that any mathematician has ever made.

JT: And before you say it, rigor is important because it alone guarantees certainty and truth.

VS: Absolutely.

JT: But be that as it may, I am interested in seeing actual deductions in progress. So far I have only seen postulates. I want to see new truths emerge.

VS: Yes, we are steadily making progress towards that goal. We have two more postulates to cover, however. Let me write out the fourth postulate:

> *Postulate 4*: All right angles equal one another.

JT: That certainly seems self-evident.

VS: It is self-evident, Judge, especially if you think in terms of a right angle being equal to 90°. But recall that Euclid only defined right angles in terms of adjacent angles that are equal. What he is saying in Postulate 4 is that any right angle is equal to any other right angle, not just the one it is adjacent to.

JT: I see your point. Euclid must have had great prescience to know what exactly to assume as a postulate.

VS: I do not think it was prescience alone. I am speculating here, but I think that Euclid probably tried out several potential postulates and finalized his list of five after a great deal of trial and error and experimentation. Along the way he seems to have heavily used the guiding principles of economy and independence.

JT: Independence?

VS: Meaning that a particular postulate is not deducible from the others. And this brings us to the problematic fifth postulate:

> *Postulate 5:* If a straight line falling on two straight lines makes the interior angles on the same side less than two right angles, the two straight lines, if produced indefinitely, meet on that side on which are the angles less than the two right angles.

JT: This one seems complicated. [*Pause*] Just in terms of length it seems to have more words than the other four put together. I am not even sure I understand it.

VS: I had to think about it for some time as well before I understood the exact phrasing. Ultimately, drawing a picture helped me a lot. I will draw it for you.

[Pause – Prisoner draws Figure 1, replicated below.]

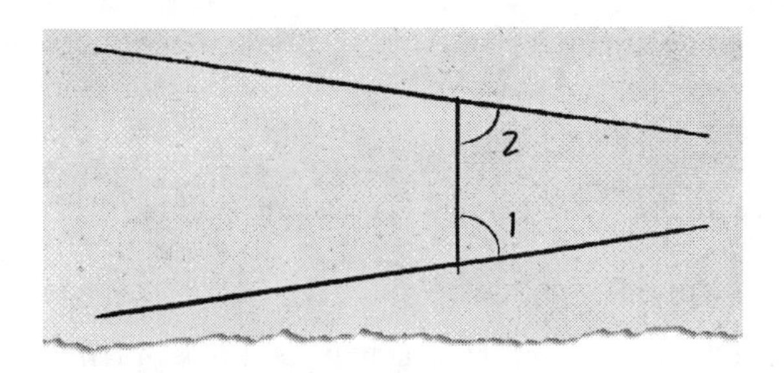

VS: I see from your eyes, Judge, that you have already grasped the meaning of Postulate 5.

JT: Yes, I have. You have drawn Angle 1 + Angle 2 to be less than two right angles, and it is apparent that the two lines will intersect on the same side as the two interior angles. [*Pause*] But I am still somewhat uneasy about this postulate. The first four postulates were self-evident in the true sense of the word. You no sooner read them than you knew that they are true. This one, on the other hand, takes a little thought.

VS: Many great minds of the last two thousand years have shared your uneasiness about this postulate. Like you, they have been annoyed by its complexity and wordiness. I have read that mankind has spent tens of thousands of hours trying to show that the fifth postulate follows from simpler postulates akin to the first four. However, these efforts have not been successful.

JT: Really! Then surely Euclid must have noticed the complexity of this postulate as well?

VS: Yes, there is compelling evidence that Euclid himself was none too happy about this postulate. For example, he refrains from using it in the first 28 of his propositions, and begins using it only when it is completely necessary to do so. Last night, as I was once again leafing through the first few pages of *The Elements*, I could vividly imagine him trying to somehow derive the fifth postulate from simpler components, and failing. Perhaps in the end he just felt in his bones that no derivations were possible and the only way to use the result was to state it as a postulate.

JT: Does this postulate bother you?

VS: Me personally? No, not really. I have read that some recent mathematicians have assumed this postulate to be false and have derived some unusual geometries as a result. But this is just vanity. I do believe that in the end this postulate is self-evident and true in an absolute sense.

JT: What do you mean "absolute sense"?

VS: I mean that it is a fundamental property of space. It is the way our universe is. If you have two lines that slant towards each other, as they must with acute interior angles, then they have to meet on the same side. It is impossible to imagine anything else.

JT: Nevertheless, as you yourself have pointed out, it is complex and feels like it needs some justification. What would have happened had Euclid not assumed this postulate?

VS: Then he would have been left with only the first 28 propositions and he would have been unable to prove important propositions that describe the nature of the world around us. It would have been artificially limiting.

JT: It still seems somewhat contrived to me. You are trying to show me facts about the universe that we can be certain of, and your certainty derives from your ability to prove statements, not just take them on faith. Yet here you are asking me to believe Postulate 5 without actually proving it. Suppose I did not believe what Postulate 5 says?

VS: But you do! And the reason you do is because it is true. It is a postulate Judge. Not everything in the world can be proved—we need some starting points. This is a starting point. Can you honestly tell me that you doubt the truth of this postulate?

JT: No, I believe it.

VS: Good. Let us proceed then. Besides his postulates, Euclid also had an additional set of axioms he described as "common notions." I will write them out for reference:

[Note from Court Reporter: Prisoner's writings reproduced below.]

C.N.1. Things which equal the same thing also equal one another.

C.N.2. If equals are added to equals, then the wholes are equal.

C.N.3. If equals are subtracted from equals, then the remainders are equal.

C.N.4. Things which coincide with one another equal one another.

C.N.5. The whole is greater than the part.

JT: I see no problem with any of these.

VS: We are then ready to move on to the actual propositions. We will prove truths about the universe that we may be absolutely certain of!

JT: About time!

VS: I realize the preparatory work has taken some time, and I thank you for your patience.

JT: It is quite all right Vijay; please do not think that I have found our conversation to be trying in any way. On the contrary, our conversations are opening up a new universe for me, and I believe that I will be able to use this type of axiomatic thinking to demonstrate matters of faith.

VS: Judge, I am very happy to hear this. It means you are beginning to see the requirements of structured thinking. But let me caution you: When you attempt to apply this type of thinking to matters of faith, you will soon see that all religious thought is quite unsupportable and untrue.

JT: That remains to be seen. Please proceed. What proposition are you demonstrating first?

VS: I will begin with Proposition 1 in *The Elements*: "It is possible to construct an equilateral triangle on a given straight line."

JT: Let me think about that. We do not have the luxury of a marked ruler or it would be straightforward. [*Pause*] Actually, it would not be quite so straightforward even then, for we will have to angle the second line so that the third line has the same length as the other two. [*Pause*] I see that this is indeed a proposition worthy of proof.

VS: Indeed it is. I shall begin by drawing a diagram.

[Pause; Prisoner draws Figure 2, replicated below.]

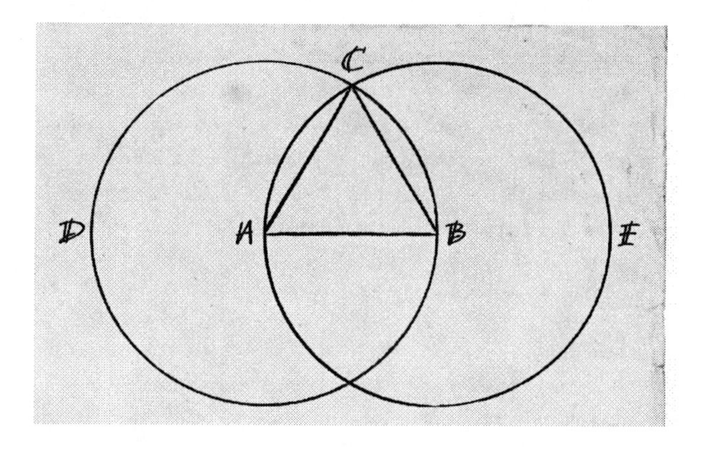

VS: Now the actual proof is quite independent of this diagram, I have drawn it only to enable us to easily understand the argument.

JT: Very well. I assume AB is the given straight line we want to construct an equilateral triangle on?

VS: Correct.

JT: I believe your figure makes the argument fairly easy to construct. Allow me this attempt.

VS: Yes, of course.

JT: First draw a circle around point A.

VS: How do you know you can do that?

JT: Postulate 3 there states that it is possible to draw a circle around a given point.

VS: Correct. Please continue.

JT: Now draw another circle around point B, the other end of the given line segment. And once again we know that we can do this using Postulate 3.

[*Pause*]

Now join the center of the first circle to the point where the circles intersect. This is the line AC.

VS: And you know that you can draw a line joining any two points by . . . [*interrupted*]

JT: By Postulate 1. I was getting to that. Now similarly draw the line BC between the center point of the second circle and the point of intersection of the two circles. Again, you know you can do this by Postulate 1. Now we are almost done, for AB equals AC as they are both radii of the circle with the center A. Also, AB equals BC as they are radii of the circle with center B. So AB = AC and AB = BC. Therefore AC = BC and all three sides of the triangle are equal and we have an equilateral triangle.

VS: That is wonderful, Judge! You have grasped the essence of the argument. This is wonderful. I only have two minor points to make: First, how do you know that two radii of a given circle are equal?

JT: That's the nature of a circle.

VS: True, Judge, but Euclid did not want us to use unstated assumptions. And we don't have to use any unstated assumption, because Euclid precisely stated the property you have used in Definition 15, where he defined a circle.

JT: I concede the point. In the interests of certainty, every step in the argument must be justified by postulates or definitions. So, yes; we know that the radii are equal by definition.

VS: Indeed. Secondly, how do you know that if AB = AC and AB = BC, then AC = BC?

JT: Come on, man! Surely that is obvious!

VS: It is obvious Judge. Even a fool would not doubt the validity of your conclusion. But we are building a proof that will lead us to certainty. Therefore we have to be extra careful. Proof insists that we state our assumptions and use them, and only them. Otherwise it is possible that we will make a mistake, and even one mistake will destroy the sanctity of proof. In this case the equality is granted by Common Notion 1 that stated "Things which equal the same thing also equal one another." Now we have completeness.

JT: The sanctity of proof? Who says you are not a religious man, Vijay?

VS: Religions require faith. Nothing we have talked about rests on blind faith.

JT: I disagree. Religion and spirituality recognize the power of the sublime. You find sublimity in your notion of truth and certainty. I find mine in the lessons of the Bible: I see you shaking your head Vijay, but let us not debate this particular issue now. We have proved Proposition 1 beyond any shadow of doubt. I am certain about Proposition 1 in the sense that you like to use the word certain. But what is your point from all this? That all knowledge needs to start with self-evident axioms and build from there? That approach may be appropriate for simple mathematical results, Mr. Sahni, but I daresay it does not apply to the complexities of real life.

VS: You believe that because you have not seen the context and scope of Euclid's work. I assure you the results get very complex very quickly. Each result builds on previous results and this process slowly but surely leads to complex results. In Proposition 2, for example, Euclid shows that if you have a given straight line and a point not on the line, then it is possible to place a line of equal length to the given line starting at the given point.

JT: That is not complex, it is obvious!

VS: Really? Please tell me how would you do it.

JT: Let me make sure I understand this. *[Note from Court Reporter: Judge Taylor drew as he spoke. His drawing is reproduced below.]* All you are saying is you have a point—let us call it A—and a line BC, that does not include A.

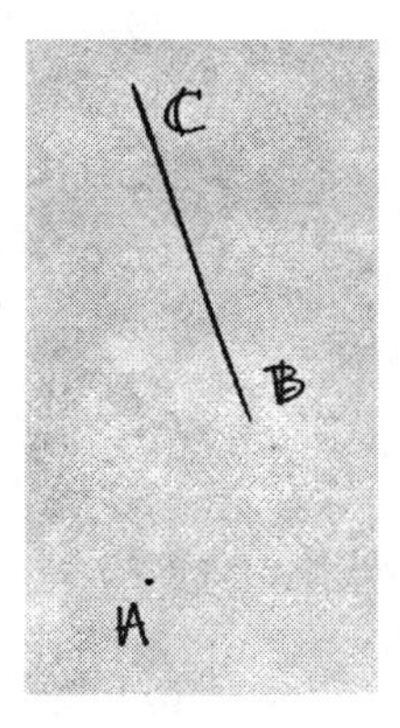

And you want to draw a line whose length is equal to BC starting at point A.

VS: Yes.

JT: All you need to do is to measure BC using a regular scale and reproduce that length on A.

VS: Which of Euclid's postulates allows use of a measuring scale?

JT: Yes, I knew you would say that. But isn't it artificial not to use the property of straight edge to measure distances? I mean, why not?

VS: Because Euclid wanted to keep his set of assumptions to a bare minimum. He wanted to assume the most basic, unassailable postulates and derive everything from them. Every new unnecessary assumption would violate his aesthetic that insisted on building his geometrical edifice upon the fewest possible assumptions.

JT: Very well. So you want me to construct the line BC on point A only using Euclid's postulates?

VS: You may also use the first proposition we just proved.

JT: You mean the one about constructing an equilateral triangle on any given line?

VS: Yes.

JT: How could an equilateral be involved in transferring lengths? Let me think about this.

[Note from Court Reporter: At this time (11:45 a.m.) Judge Taylor ceased conversation. Subsequent to this he occasionally spoke to himself and drew various geometric drawings that were not satisfactory to him. Neither of these are part of the Court Transcript. He also frequently referred to a list of definitions and postulates that the prisoner had hand-written on a sheet of paper. At 12.35 p.m. Judge Taylor resumed conversation.]

JT: Mr. Sahni, I am thought in my circles to be an intelligent man. But I confess I have made no progress in this transference of length problem. Euclid's postulates seem too barren to be able to do this. As you can see I have drawn circles and equilateral triangles but to no avail. Please show me the solution.

VS: I would be happy to show it to you, Judge. And please do not berate yourself for not seeing the construction. When I first read *The Elements* I tried to prove each proposition for myself without referring to Euclid's solutions and I vividly recall that I had some difficulty with this proposition. Anyway let us begin by referring back to your drawing.

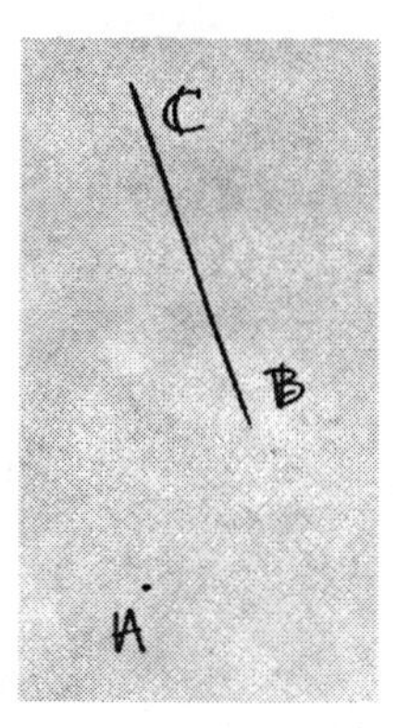

We want to draw the length BC on the point A. So let us begin by drawing a line between point A and point B:

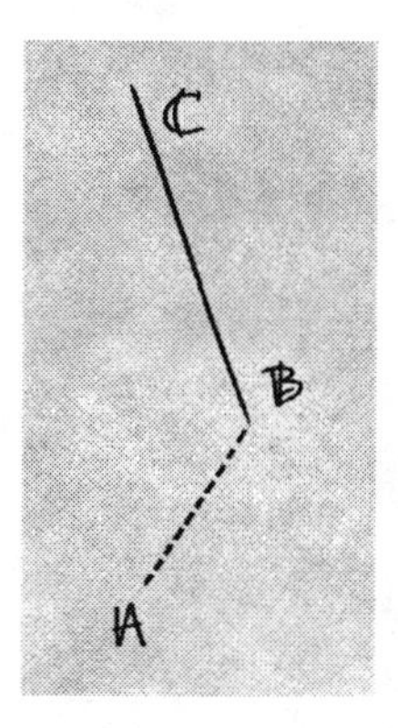

JT: You know you can do this by Postulate 1. It was actually the first thing I did.

VS: Good. Postulate 1, as you know, allows us to draw a line between any two points and that is exactly what I did. Next I want to draw an equilateral triangle on the line AB. Here is what my construction would look like:

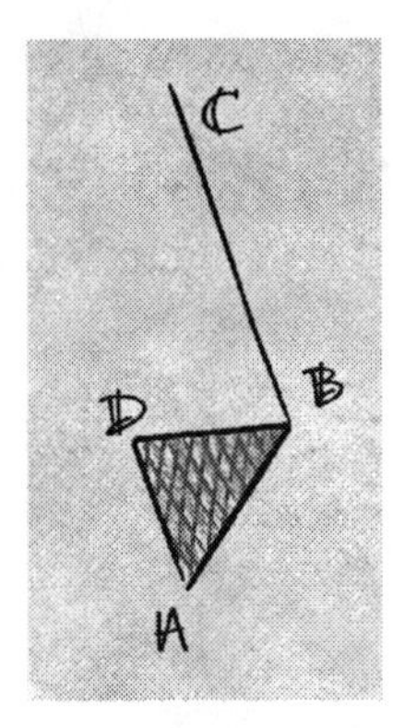

JT: There is the equilateral triangle from Proposition 1!

VS: Precisely.

JT: But I noticed you didn't draw the two circles that we used in the proof of Proposition 1.

VS: Those would just interfere with our current construction. We are using the result of Proposition 1 without redoing the construction.

JT: I had even tried this step in my drawings, but did not know what to do with it.

VS: Take a look at this:

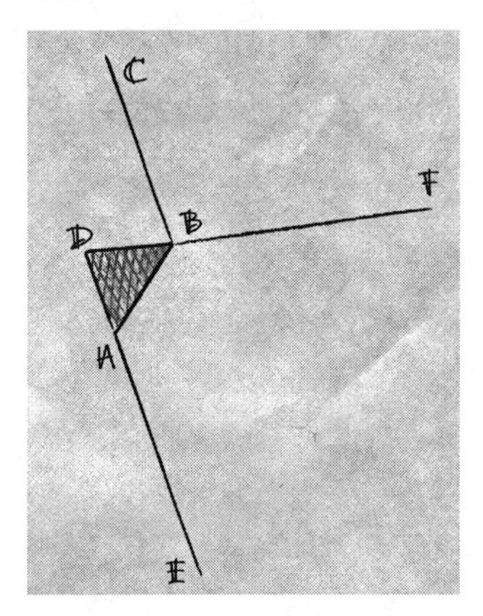

JT: What happened here? You extended the sides DA and DB?

VS: Yes. What gives me the right to do this?

JT: Let's see. That would be . . . Postulate 2: It is possible to produce a finite straight line continuously in a straight line. But why did Euclid do this?

VS: You will soon see! Take a look at this. Postulate 3 gives me the ability to describe a circle with any center and radius. I choose B as the center and BC as the radius and draw a circle.

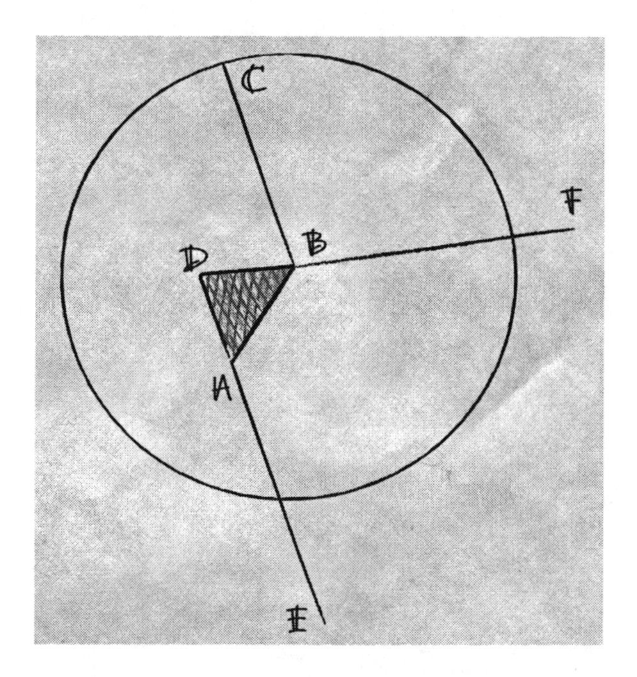

JT: Okay, but so far there is no line on A whose length is equal to BC.

VS: You are correct. AE is in fact an extension of any length and is not equal to BC.

JT: Right. All we know is that AE could be extended longer than BC.

VS: Absolutely. But now let's mark off the points G and H where BF and AE intersect the circle. Here is the same drawing with the points G and H shown:

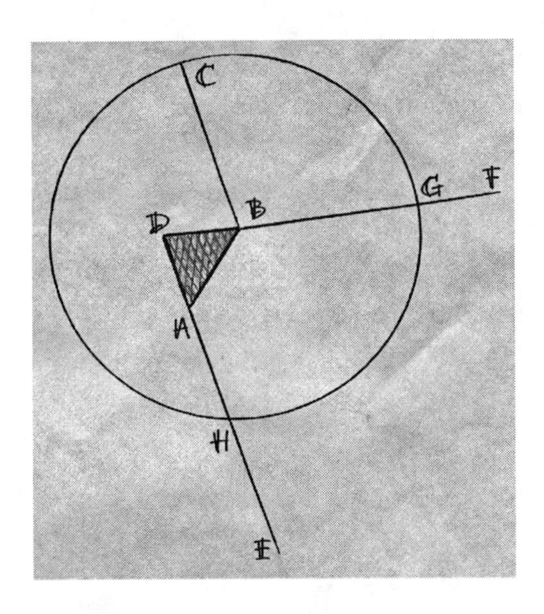

JT: I feel another circle coming up.

VS: Good intuition Judge! You are correct. I now draw a circle with center D and radius DG. Here is what I get:

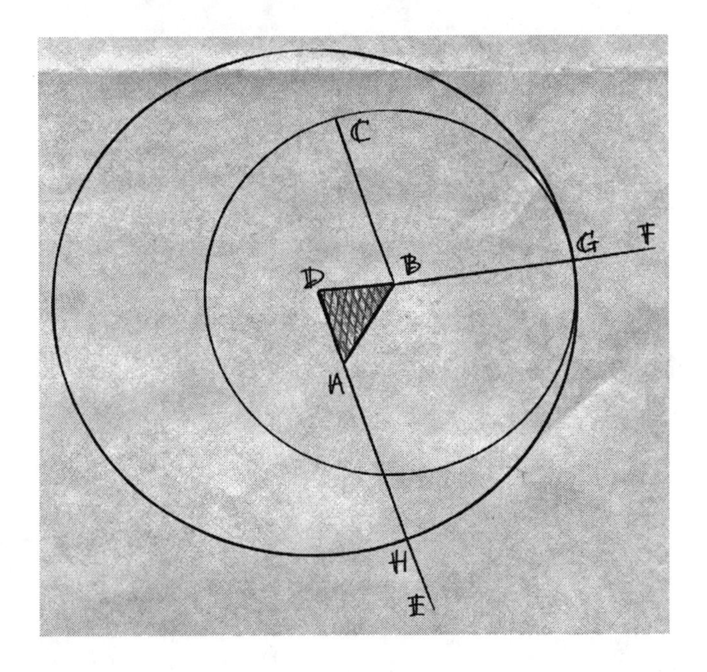

JT: Interesting. Let me look at this. We have the small circle with center B and radius BC. BG is also a radius of the small circle, so BC = BG.

VS: How do you know that?

JT: By the way Euclid defined a circle. Now in the big circle we have DH and DG as the radii, so DH = DG. *[Pause]* But wait a second! DA = DB because they are sides of an equilateral triangle. So DH – DA = DG – DB.

VS: What gives you the right to subtract lengths like that?

JT: Why, that would be Common Notion 3, which said, "If equals are subtracted from equals, then the remainders are equal." But DH – DA is precisely AH, and DG – DB is precisely BG. So AH is equal to BG. Aha! Now BG and BC were equal because they are radii of the smaller circle. So AH = BC, which gives us a line of length BC on the point A. That is magnificent! It is beautiful.

VS: Yes, it is beautiful. More importantly it is certain. Certain to the last detail.

JT: It is certain Vijay. I give you that.

[Note from Court Reporter: Conversation ceased for four minutes.]

VS: What are you thinking Judge?

JT: A number of things, Mr. Sahni, but mostly I was thinking about this passion for certainty you clearly have and Euclid clearly had, and I'm wondering why I never had it and why you feel such a strong need for it.

VS: I feel a strong need for it, Judge, because without certainty there is no truth; there is only guesswork. And guesswork is not something to base a life upon. I maintain that the only way to certainty is the axiomatic method where we begin with truths that are apparent to any thinking mind and use our logic to build new truths. Any knowledge not using these methods is flawed.

JT: Why is it flawed?

VS: Because unless we enforce the discipline of axiomatic thinking our human brains fall prey to what I call "upside-down reasoning."

This is reasoning in which the end result precedes the supporting arguments. People make up their minds first, and tailor their arguments to support the position they want to take.

JT: In your mind religion is an instance of this type of thinking?

VS: Yes it is. People are driven by emotional or cultural imperatives to accept religion and then they find justifying arguments to make their faith appear more reasonable. But these arguments are never axiomatic and precise. And this process is not just limited to religion. In every facet of life I believe that the human brain is conditioned to seek the end state first and find the justifications afterwards. The problem is that this type of thinking is prone to error because it derives from ephemeral and subjective wants. Therefore I will repeat what I have always said: the only way is the way of logic—logic founded on unassailable postulates.

[Note from Court Reporter: Conversation ceased for several minutes.]

JT: I understand what you have said. I understand the case you are trying to make. I need to give myself time to take this all in, but I see it is well past the time for me to leave. Do you want to tell me anything else before I begin deliberations on my decision?

VS: No, Judge, I have nothing new to say. I have shown you the essence of Euclid. From here the demonstrations get more complex, but their nature is the same as what you have just seen. There are long chains of logic, each new result building on the ones before it, each as certain as the postulates we started from. And this is the way of knowledge: all knowledge that is worth believing and relying upon must be based on reason alone. If you agree with this you must recommend my acquittal. If you don't, then I am ready to go to trial, and given the sentiment in these parts, I would best be ready for a long sentence as well.

JT: That remains to be seen. I can only promise that my decision will be based only on the application of the law to the best of my judgment and ability.

VS: I believe that. OK, then. You must leave now?

JT: Yes, Mr. Sahni, but first let me shake your hand. These conversations of ours have often been difficult; I have found many of your views to be alien to my way of thinking. But you certainly have awakened me to the pleasures of geometry, of this I am sure. Here is an entire subject for me to focus on and enjoy—a new universe, really. Its rigors suit my mind and solitary nature and I know that I will be spending many an evening with Euclid's *Elements*. Whether the methods of geometry can and should be applied to all knowledge is the question I am required to consider, but regardless of my answer I know that I am coming out a better man for having met you. I thank you for the gift of geometry.

VS: Judge, you are quite welcome. You have a keen mind indeed, and it has been liberating discussing these issues with you. My only wish is that our encounter had occurred in less unusual circumstances. Perhaps someday we will meet in the Punjab, and I can greet you as a free man.

JT: Perhaps. Farewell.

• • •

There is a graveyard on a tree-lined street near the Stanford foothills which I usually avoid including in my running route because it involves climbing up a steep slope that crests right at the cemetery's gate. But since running with Claire had made me fitter—or so I hoped—I decided to test myself against the gradient just to see if my lungs could handle the workload any better than they could a year ago, the last time I had come this way while running.

The scamper up was easier than it used to be, but it still was not easy. The late morning sun was hot and my speed was more ambitious than it should have been. So by the time I made it up the hill I needed to rest a minute. And that was when I saw Nico. The lanky frame and the black leather jacket made me suspect that it was him, and the motorcycle helmet hooked in his left arm made me sure. He stood tall and straight, his head bowed slightly forward, presumably looking down on a gravestone. What struck me first was his absolute motionlessness. He could have been one of the statues scattered throughout the cemetery except for his long, white hair, which I could see being swept around with every gust of wind.

I watched him there, not quite understanding why I was doing so. Five minutes stretched to ten, which lengthened into the afternoon. The street was completely soundless, and I remember noting the absence of bird or squirrel sounds, which struck me as odd given the number of nut-bearing oaks around. Looking at Nico, I began to think of Bauji and his cremation. My parents and I had taken his ashes for dispersal in the Ganges, and I flashed on being in a boat under a railway bridge slowly pouring his last material remains into the river. The water had been perfectly still in parts and seemed flowing in others. Out of nowhere some ducks appeared, and they had floated a few feet away, silently looking at us. Just as I emptied out the container a train had passed overhead. And then it was over.

Nico, as was his wont, greeted me with a warm smile when he came out.

"Ravi! What are you doing here?"

"I was out on a run and saw you , so I stopped." I replied.

He smiled. "And you're probably wondering . . ."

I interrupted him. "No, no. I was just surprised to see you standing there so still and, at least seemingly, at peace. I mean, you're so restless in the classroom with all that pacing," I said. Nico always made it easy to say what one was really thinking.

He laughed. "Years of practice. My wife is buried here. She died 20 years ago in a motorcycle crash, and I've visited her ever since."

"How often do you come?" I asked.

"At least once a week, sometimes more."

"And what do you do?"

"I pray," he replied, looking into my eyes.

This was entirely a new side of Nico. Through all our discussions I had just assumed that he was secular like most others I knew. My surprise must have shown on my face, for Nico was laughing again.

"Why are you so surprised that I pray?" he asked.

"I had no idea you were religious," I said.

"I'm not religious in the sense that I have deep allegiance to a particular religion, but I am religious in the sense that I have faith and I believe in God." He spoke easily without feeling the need to be defensive.

I wondered if Nico thought my grandfather to be immature and misguided. He had read each one of the transcripts, and it had never occurred to me that he might share the judge's point of view over Bauji's.

As he did many times in class, Nico anticipated what I was thinking. "Ravi, your grandfather was a good man. He was questioning and questing and there is nothing wrong with that. In fact, I admire him for it. I don't need to judge other people's faith or lack of it," he said, and I knew he meant it.

"What do you pray for, Nico? When you stand there near your wife's stone, what's in your head?" I really wanted to know.

Nico nodded and his eyes narrowed in thought. "I try not to think of anything," he said. "I stand there, usually with my eyes shut, and I try to visualize her face, and I pray that she's happy and at peace."

"So you believe in the afterlife?"

He laughed. "I don't know," and from the lower pitch of his voice I got that he didn't really want to debate his beliefs with me.

"Can I give you a ride or are you going to finish your run?" He asked patting the back seat of his motorcycle.

"I'll take the ride," I said.

• • •

BARUCH SPINOZA JOURNAL ENTRY, 1656

Today I was excommunicated.

I was excommunicated for speaking the truth. It seems the truth was too stark for those who sat in judgment over me. They could not bear to hear that nowhere in the scripture does it say that angels exist or that the soul is immortal. To me these are facts, and there is a clear proof of these facts; you only have to read the scriptures and you will be convinced. But these excommunicators of mine seem to believe that one may not apply logic to religion. Instead of being persuaded by my common-sense deductions, they have accused me of blasphemy and thrown me out of their society.

Much to my surprise, I am not too distressed. Instead I find myself somewhat relieved. I am now freer than I have ever been before—free to think and write in the manner of my choosing. And I will choose to focus on finding the truth about things rather than repeating spurious ideas without rational examination.

It seems to me that rational understanding of anything consists in seeing it as the logical consequence of its cause, just as the properties of Euclid's geometrical figures are understood by seeing them as the logical consequences of relevant definitions and postulates. I am more and more convinced that I should attempt to lay out an entire philosophical system in the manner of Euclid: I will begin with some definitions, follow them with some well-chosen self-evident postulates, and build the results of my philosophy one step at a time just as Euclid so patiently did for his propositions. This way I will construct a philosophy of ethics and metaphysics that will be as certain as Euclid's geometry. After all, why should only the mathematicians have the keys to certainty?

• • •

DAVID HILBERT JOURNAL ENTRY, 1885

Sometimes people surprise you. For my birthday last week my sister gave me The Essential Spinoza, *which was a compilation of the philosopher's most famous essays. I told her that this was not the type of book I read and that she should best return it for a refund. But she shook her head. "I know you'll like it. Give it a chance," she had said confidently.*

Later that night, despite myself, I opened the book and began to read and found that my sister knows aspects of me better than I do. I found myself deeply stimulated by a subject other than mathematics. I didn't realize that was still possible.

It turns out that Spinoza tried to derive conclusions about life from a set of axioms just the way that Euclid derived geometric theorems from his five axioms. I think Spinoza was not completely successful—partly because he could not conform to the discipline of thought that Euclid's ways require, and partly because his subject is quite difficult. Life, after all, is even more subtle than Euclid's geometry!

But Spinoza's failure got me thinking: What should we require from an axiom set? I've only thought about this for a few hours but it seems to me that at the very least an axiom set should be correct, meaning that it should not lead to a contradiction. I deeply believe that mathematics (or life, for that matter) does not permit contradictions and an axiom set that leads to a contradiction must be flawed.

Say I have two axioms: (1) *God exists and* (2) *Everything that exists occupies one place at one time. Let us also say that by definition God is a being that is omnipresent. Well, I'm sure you can quickly see that I would have a contradiction on my hands. If God exists, then an omnipresent being exists, which contradicts the constraints of my second axiom . . . which means that my axiom system is not correct.*

It is not hard to find axiom systems that are correct. I feel quite sure that the axiom systems underlying the laws of numbers, for example, are correct. Euclid's axioms also seem to be correct. After all, they have not yielded a contradiction in over 2000 years of rigorous testing!

Leaving correctness aside, my second objective from an axiom system is that it be complete, meaning that all theorems may be derived from that set of axioms. Would it not be marvelous for a single set of axioms to explain every single theorem in mathematics?

Better yet, would it not be a wonderful thing to have a complete and correct axiom system for all of life?! We would be able to demonstrate everything with complete certainty. There would be no philosophical quagmires and there would be no room for this persistent existential purposelessness that seems to infect every intelligent person I meet. I wonder, if such an axiom set exists, would we be able to know it?

Wir müssen wissen, wir werden wissen—We must know, we shall know.

• • •

It turned out that Adin had stopped participating in Thursday Night Jazz because he had landed a gig at a café in San Francisco. Apparently he had been a regular in a Jazz band (The Swinging Sequoias) and finally felt confident enough to invite people to come and hear him. His e-mail summons seemed unlike him: the taciturn precision that I had become accustomed to had been replaced by a more emotional appeal.

From: Adin Kaminker
To: Ravi Kapoor, Peter Cage, Percy Klug, Claire Stern
Hey, Infinity Gang!
I'm a very fortunate saxophone player, to be able to play jazz with Ray Davis on the piano, Jim Greene on drums, and Lil' Jo Harrison on bass. These guys are capable of *anything*, and if you want to hear it live, come on down to "Simple Pleasures" on Thursday at 9:00 p.m. (Balboa & 35th in San Francisco). This is some seriously exhilarating music we're playing.

People don't "go out" as much as they used to. Maybe access to a wide array of entertainment on TV contributes to people staying home, and that's led to a society of people who have a hard time putting down the remote. It's important to realize, though, that most art, whether it's a painting in a museum, live theater, or mathematics, demands in-person participation and support. You, as a potential audience member, make the decision as to whether art can survive, by simply choosing whether or not to attend an event.
I hope you come,
Adin

Simple Pleasures turned out to be a comfortable café on a fogged-in street on the western edge of San Francisco. I drove there with Claire and PK (Peter had a concert to get to later that night, so he drove separately). This was the first time I had seen Claire get dressed up in anything other than her customary black top and jeans. She had put on a black dress and I thought I even detected a touch of make-up. She looked fine.

After getting coffee and hot chocolate, we moved to the back room where the music was coming from. There were people sitting on old sofas and on the Persian rug on the floor. Many had books open, others talked, and a couple people sat perfectly still with their eyes shut. Hardly anyone seemed entertained or bothered by the music that Adin and three others played from the corner near the fireplace. Peter was already there in his neatly pressed khakis. PK found a spot on the carpet near Peter's couch and seemed, as he did everywhere, totally at ease.

Adin on the saxophone was terrible. His slender frame crouched over the instrument, giving me the impression of an old lady at the wheel of a big car. He missed notes entirely, he was frequently late and sometimes early, and he could not keep up with the tempo set by the drummer. Nevertheless, he seemed happy. Every time he would make a mistake he would shake his head and apologetically smile at his mates, who did not seem to be too bothered by his musical shortcomings. They were more concerned with their own solo pieces which, relative to Adin, they performed with varying degrees of practiced smoothness. The piano player was far and away the talent of the band. His sense of timing lived inside the muscles of his fingers, and he played without the conscious effort that burdened Adin. His tone, especially on the "Blue" of "Blue Moon," were quietly soulful. But again and again, my eyes went back to Adin. His playing, despite its apparent deficiencies, was communicative, enthusiastic, and joyful. He seemed to be completely different from the reticent philosopher who parsed Bauji's transcripts with me.

After a slower, brooding rendition of "The Way You Look Tonight" and another piece that I had heard but couldn't place, the band announced a 20-minute break. PK clapped loudly and waved Adin over.

"What did you think?" he asked us, looking at no one.

"Hey, hey!" said PK jumping up for a two-handed high five, to which Adin responded uncertainly, unsure if he could believe any congratulations sent his way. "Dude, that was cool!" said PK, giving Adin a bear hug. Claire, unwilling to lie as directly, added a noncommittal "Uh-huh." Peter leaned further back on the sofa, broadcasting that he was not about to volunteer anything. And so Adin looked at me with worry and hope, and an unspoken question.

What to emphasize, kindness or truth?

"Adin, that was interesting, but I think you're a better philosopher than musician," I told him.

Adin laughed, and as I had hoped, the compliment had blunted the implied criticism. He nodded wistfully, "I wish that wasn't true, but I'm afraid it is."

Over drinks and between more rounds of music, we read the pages of Bauji's latest transcript. Claire was fascinated by the way Euclid had taken the axioms to build the construction that transferred lengths. PK couldn't believe how complicated the whole process was and what one had to do to prove Proposition 2. "Why not just measure the line on a straight edge and draw it wherever you want?" he asked, to which Adin shook his head vigorously but didn't look up and interrupt his reading. Peter was not that interested in the whole idea of certainty—he felt certain enough about most things without having to question the source or validity of his belief. He got involved in a rapid-fire chess match with a young Russian who was playing for $5 a game. Peter lost the first two, but took the third.

Adin's musician persona ebbed with each page of the transcript. He was quiet as he read, oblivious to his surroundings. His band partners were ready to resume playing and the bass player came over to our table to get him, but Adin shook him off. He read long after the others had finished and then finally he said, "This is the kind of analysis I've been looking for forever."

The band started playing without Adin, but he didn't notice. He had switched. "Human beings have a great propensity to decide things based on how they feel. What we buy is influenced by advertising, our political leanings are determined by cultural upbringing, just about every opinion we have is a product of some nonrational process. It is an accident based on a momentary whim or the opinions of those around us in childhood. Hardly anyone examines things deductively; people are satisfied with the feeling of being right, rather than following the rigor required by certainty. Your grandfather spotted this and he couldn't tolerate it. He talked about it in the context of religion, but it's rampant in all spheres of human thought." After slouching for the last several minutes, Adin suddenly straightened up. "The thing is that we humans are not condemned to this state of affairs. The good news is that we can analyze things, reduce them to first principles, examine our assumptions, and live sensibly, rationally and without contradictions. This is what your grandfather was trying to get at."

I felt like an outsider, a spectator bridging my grandfather and his rightful intellectual heir. To me Bauji had passed on his love for mathematics and the ability to appreciate its beauty, but for reasons known only to him he had kept the philosophical implications of axiomatic mathematics very much to himself.

Suddenly, and without apparent reason (at least the reason was not apparent to me then), I got irritated with Adin. "So, if this axiomatization thing is so great how come philosophers have not tried to use it in real life? Why is it limited to mathematics only?"

He shook his head. "Philosophers have tried to extend it to the philosophical realm. Spinoza tried it, Descartes tried, and they are still considered to be thinkers of the highest order."

"Yes, but Adin," I said giving him my best "get real" look, "life is too complex, it has too many shades of grey. People couldn't possibly make important decisions in their lives by going through some purely deductive process."

"Why not?"

"I'll tell you why not," I said. "Here's a real-life example from my life. Yesterday I was offered a second-round interview with Goldman Sachs. But the problem is . . ."

"Ravi, you got the call from Goldman?" interrupted Peter. "Dude, that is totally fantastic!" he said excitedly, even ignoring the precious seconds passing on the chess clock beside him.

"Thanks," I said. "The problem is that I feel really conflicted about the offer. I've been anticipating it for many days and my feelings about it have alternated just about every day. Sometimes I think of how happy the offer would make my parents, of how proud they would be, of the huge fuss they'll make, of how welcome the money would be, and of all the things the money would solve."

"And other times?" Surprisingly it was Claire who asked this, not Adin.

"Other times I feel tired just thinking about going and working at a bank every day."

This last bit offended Peter, who gave up on the chess match and explained why investment banking had a bad rap but was really a thrill a minute. "Many people try to get into this profession, Ravi. You should feel fortunate to get this far." I told him I intellectually understood what he was saying; I even agreed with it, but I couldn't help what I felt sometimes.

PK recommended a dose of self-examination. He said that I should try and analyze myself to see which of my two alternating and opposite feelings was more real. I told him that if I could do such an analysis there would be no problem at all.

Adin got us back on track. "Ravi, you claim that this decision you're facing—to pursue this job offer or not—is outside the reach of deductive reasoning. You're wrong."

If anyone else had been this abrupt it would have surely been offensive, but Adin said it so matter-of-factly that it came out as an innocuous observation, as if he was saying that it was raining outside. "I'm a Jew," he continued, "and my religion has hundreds if not thousands of written rules on how human beings should behave. Now these rules are nothing but axioms that are interpreted and applied with the help of rabbis. The rabbis take what is in the scriptures and help people make decisions exactly like the one you are facing. At its core, Judaism is a deductive religion. Now, you may disagree with some of its axioms, but for Jews that do agree with the principles in the Torah, there is a reliable way to make decisions in life. And they have certainty about these decisions."

The crowd in the café had thinned. The other three members of the Swinging Sequoias were packing up their gear; but Adin was warming up to a different type of riff that had Claire, PK, Peter, and me rapt: "The U.S. Constitution is also nothing but a set of axioms. In fact, the framers even say. 'We hold these truths to be self-evident'. Self-evident truths are another name for axioms. The axioms of the Constitution are interpreted by judges. So in its ideal state the laws of this land have deductive certainty."

PK took exception. "But people disagree about the laws. How can you say they are certain?"

"I know people disagree about the laws, and it's for one of two reasons: either they disagree with the axiom or they disagree with the interpretation. The beauty of Euclid's methods is that he allowed disagreement on neither score. His axioms are completely self-evident and the methods of interpretation are rock solid. Therefore, Euclid provides certainty. So if Euclid could guarantee certainty in geometry, who is to say that similar certainty is not possible for intractable-looking decisions, such as whether Ravi should pursue Goldman?"

Who is to say indeed. Perhaps Adin was right. Perhaps our purpose in life was to find a set of absolutely true axioms and to make all our decisions based on the deductive interpretation of those axioms.

With a start, Peter realized he was late for his concert and PK asked if he could go with him. Adin already had a ride, so Claire and I drove home alone. But the sparkling conversation I was hoping for didn't materialize.

Claire looked out the window while I steered us through the thick fog that had rolled over the Richmond dune. We drove through Golden Gate Park in silence, and it wasn't until we were on the freeway that she spoke. "So you're really thinking of getting into investment banking?"

"Maybe," I allowed.

She nodded and looked away again. And then, with a force of will, she set aside whatever was bugging her. "You know Ravi, I've been thinking about a problem. You remember how Nico showed that the infinity of real numbers between 0 and 1 is greater than the natural numbers?"

"I don't think I'll ever forget that proof, Claire," I had said. And as the years have gone by I have indeed not forgotten. I can still reconstruct that proof in my sleep.

"Now let me ask you this. Is there an infinity greater than that of the real numbers?"

Interesting question. I had not thought about it. An infinity greater than the reals—what could it look like?

"It took many hours and finally I think I saw it," said Claire as she smiled, a fast, happy upturning of the corners of her lips that was shy and proud at the same time. "I just considered the points on a two-dimensional plane."

She described her construction quickly, almost breathlessly. At one point she took out her notebook from her purse and drew out her idea, which I examined from the corner of my eyes, trying to ensure I kept the car between the white lines. Claire's argument was simple and intuitive:

Consider all the real numbers between 0 and 1. You can draw them on a line, where each point represents a real number:

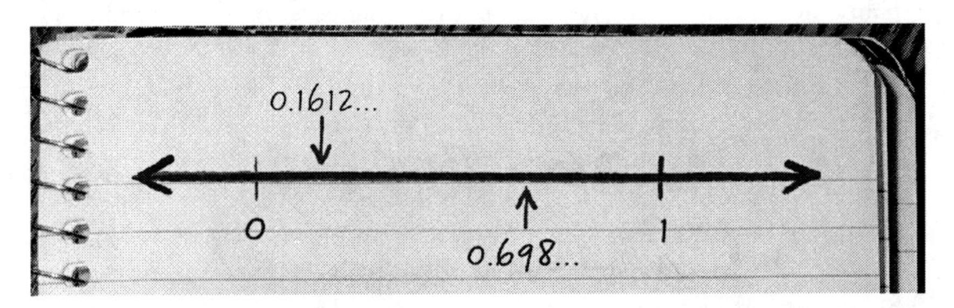

Intuitively it is clear that the real number 0.1612 . . . (recall that real numbers are allowed to have decimal representations that go on for ever and do not repeat) represents a point somewhere between 0.16 and 0.17. The real number 0.698 . . . occurs to the right, closer to 1.

Claire's idea was to consider all the points on a plane and not just on a line. She did this by considering the space defined by two lines. One line was horizontal (like the one above) and the other she turned around on its head to be vertical, perpendicular to the original line. The two lines intersected at their respective 0 points.

Every point on the plane could be named by two reals—one from the horizontal line and the other from the vertical. On her drawing Claire marked a point on the plane that corresponded to the 0.741 . . . point on the horizontal line and the 0.351 . . . point on the vertical line. She marked the point as (0.741 . . ., 0.351 . . .). The point on the top right corner was (1, 1) and the point on the bottom left was (0, 0).

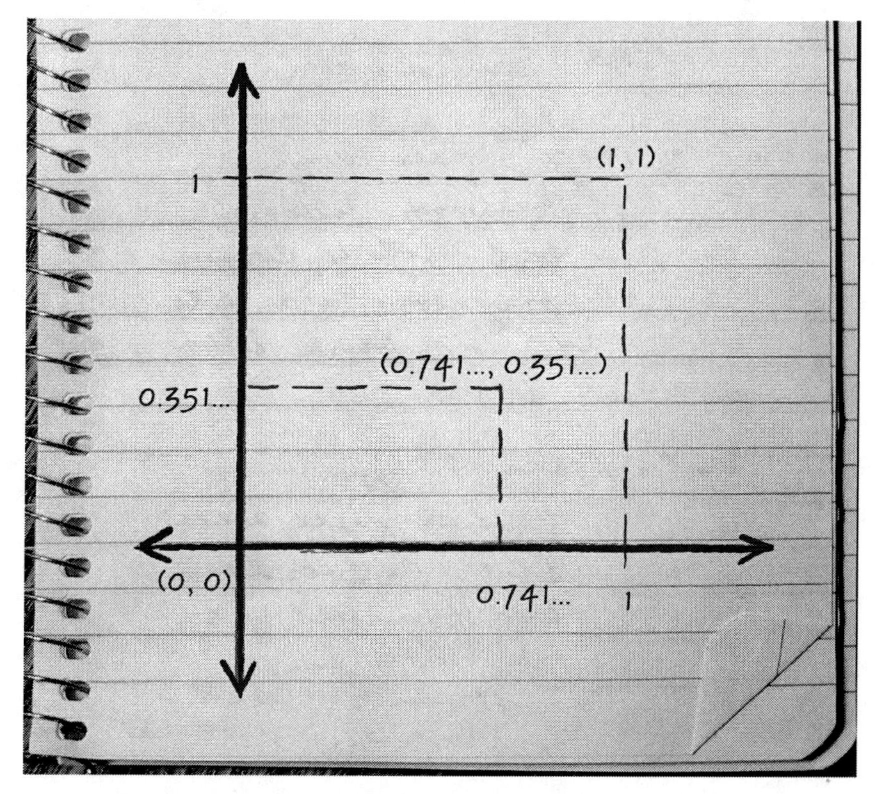

In general, each point could be thought of as (a, b), where a and b are real numbers. Claire's contention was that the infinity of the points on the plane was greater than the infinity of points on the line.

This was a completely plausible idea, and I could immediately think of a way to extend it. "And, Claire, if you drew out a cube where each point in the space of the cube is represented by (a, b, c) then you'll probably get an infinity that is greater than the infinity of points on this plane."

She was with me. "You're right! And the four-dimensional space will be greater than the three-dimensional space even though we can't visualize it. Then we'll have successively greater infinities with each new dimension! An infinity of ever larger infinities!"

It seemed so natural, it had to be true. We were thrilled to be able to see it. Suddenly, the earlier distance between us was gone and everything seemed right. When I dropped her off at her apartment, Claire gave me a hug. "It's wonderful to be able to share beautiful ideas."

But not all beautiful ideas are correct. When I got home I tried to show that a contradiction must arise if there is a one-to-one map between the set of real points in (0,1) and the set of real pairs that constitute the unit plane. Hours went by and my hoped-for contradiction never emerged. It was about 3:30 a.m. when I realized that a contradiction could never emerge *because there was a mapping from point pairs in the plain to points on the line*!

Say you have a point pair on the plane. Call it (a, b). Further, let's say the decimal expansion of a and b is as follows:

$$a = 0.a_1a_2a_3a_4a_5a_6 \ldots,$$
$$b = 0.b_1b_2b_3b_4b_5b_6 \ldots$$

Now we want to construct a mapping that maps this point pair to a *unique* point on the real line.

Here's the brainwave: Construct the point on the real line (let's agree to name it z) by alternating the digits in the decimal expansion of a and b.

$$z = 0.\ a_1b_1a_2b_2a_3b_3 \ldots$$

For example, our earlier point pair would map as follows:

$$a = 0.741 \ldots,$$
$$b = 0.351 \ldots.$$

Then alternate digits to construct z:

$$z = 0.734511 \ldots.$$

Notice that each point (a, b) on the plane is uniquely associated with a single point z on the line! So despite the difference in dimension, the points on the plane are no more abundant than the points on a square!

I found out the next morning from Nico that Cantor had had the same intuition that Claire did in constructing the potential hierarchy of infinites and then found the same proof I did to show that his intuition was incorrect. My proof, Nico said, was exactly the one Cantor had used, save for some technicalities that arose because of the nonunique decimal expansion of real numbers. Even going through a three- or four-dimensional cube didn't change things; they all had the same number of points as the real line. Needless to say, he was surprised. It must have seemed to him then that it is impossible to exceed the infinity of the line, but he himself was to find that this was not true. There is a way to build ever larger levels of infinity.

Even though it was 4:00 a.m. I called Claire to tell her what I'd found. She was sleeping but I could hear her smile when she realized it was me. I explained my mapping and she got it right away. "That's hard to believe, but I see your construction, so I have to accept it," she had said, echoing Cantor.

Then, we talked until dawn. She told me about her childhood, about how she got interested in mathematics (her fourth-grade teacher), about how difficult it had been for the allegedly enlightened establishment to take a woman mathematician seriously, about how she saw the rest of her life unfolding, and even about the kind of man she wanted to be with. "Someone like you," she had said, laughing.

I repeatedly replayed that line in my head after we had hung up. Just before drifting into sleep it occurred to me, for the first time in my life, that death was indeed something to fear.

• • •

Nico began the next class by announcing that he was going to talk about one of the "deepest and most celebrated problems" in all of mathematics. He referred to it as the Continuum Problem. "Before we get to the actual problem we need to do some groundwork. I'm going to begin by asking a question that certainly seems innocent enough. We saw last week that the infinity of real numbers is greater than the infinity of natural numbers. Is there an infinity greater than the infinity of real numbers?"

I caught Claire's eyes and she smiled back in recognition, but Nico misinterpreted my look; he thought we were not paying attention. "Ravi, you look pretty smug up there. Do you happen to know the answer?" He was mollified (and amused) when I told him about how Claire and I had thought the infinity of the set of all pairs of real numbers might fit the bill but had gone on to convince ourselves that it was actually the same size as the infinity of real numbers. About our method of alternating digits he said,

"It's great that you guys hit upon the correct approach. About a hundred years ago Cantor had asked himself the same question and on finding the proof he wrote to a friend, 'I see it but I don't believe it!'"

I completely understood what Cantor must have meant. Once again infinity had demonstrated the fallibility of our intuitions.

To show an infinity greater than the infinity of real numbers and to lay the groundwork for the Continuum Problem, Nico introduced two new concepts: cardinality and power sets. Cardinality is easy enough to understand—it is simply the number of elements in a set. Nico had told us in the very first class that a set is "a collection of objects." So the cardinality of a set is a count of the number of objects in the set. For example, the cardinality of the set {a, b, c} is 3.

Nico wrote out some other sets and their cardinalities:

Set	*Cardinality*
{elephant, hammer, 45, q}	4
{Carol}	1
{United States, India, Israel}	3
{1, 2, 3, 4, 5, 6, 7}	7
{}	0

Nico referred to the last set in the table, {}, as the empty set. It contains no elements; hence its cardinality is 0.

Then Nico considered the infinite sets. The cardinality of the set of positive integers is infinity. Likewise the cardinality of the real numbers is also infinity. But Cantor had shown that they were different levels of infinity, so he gave them different names.

Set	*Cardinality*
Positive integers = {1, 2, 3, 4, 5, 6, 7, 8,. . .}	$\aleph_0$
Real numbers	c

ℵ is the first letter of the Hebrew alphabet. Nico pronounced it "Aleph." The "0" subscript referred to ℵ being the first cardinal in a series of infinite cardinals. A cardinal is simply a number that denotes the cardinality of a set. In this sense the numbers 7 or 4 or 451 are all cardinal numbers, as there are sets with 7, 4, and 451 objects. The cardinality of the real numbers is denoted by "c" for the continuum of the reals on the number line. Cantor's diagonal argument had shown that $c > \aleph_0$. And the fact that there are as many rational numbers as natural numbers meant that the cardinality of rational numbers is $\aleph_0$.

Power sets turned out to be equally straightforward. A power set is simply the set of all subsets of that particular set. As usual Nico explained by using examples:

The power set of {a, b, c} is {{a}, {b}, {c}, {a, b}, {a, c}, {b, c}, {a, b, c}, {}}. Each element is a subset of {a, b, c}. In particular, {a} is a subset of {a, b, c} and therefore {a} is a member of the power set of {a, b, c}. Nico also pointed out that the empty set {} is a subset of all sets and therefore is a member of the Power Set of every set.

"Why is that?" PK asked.

"To show a set is a subset of another set one must show that each of its elements belong to the other set. Well, an empty set has no elements and is therefore vacuously a subset of every set."

Just to make sure everyone had internalized the idea of a power set, Nico provided more examples:

Set	*Power Set*	*Cardinality of Power Set*
{*, $}	{{*}, {$}, {*, $}, {}}	4
{1, 2, 3, 4}	{{1}, {2}, {3}, {4}, {1, 2}, {1, 3}, {1, 4}, {2, 3}, {2, 4}, {3, 4}, {1, 2, 3}, {1, 3, 4}, {1, 2, 4}, {2, 3, 4}, {1, 2, 3, 4}, {}}	16
{}	{{}}	1

Nico asked if anyone saw a relationship between the cardinality of a set and the cardinality of its power set. "If a set has five elements, what is the cardinality of its power set?" he asked.

Claire had the answer almost before Nico finished asking the question. "32," she replied and then for good measure generalized her result. "If a set has cardinality n, then its power set has cardinality 2^n."

As always, Nico wanted to push us another step. "Proof?" he demanded.

Claire was ready for him. She had figured this out in lightning speed. "Each subset is defined by whether a particular element is included in it or not. So, for each element we get two types of subsets: one that contains the element and the other that does not. So, if there are two elements, you get 2×2 subsets; for three elements you get $2 \times 2 \times 2$ or 2^3 subsets; likewise for n elements you'd get 2^n subsets."

Many in the class turned to look back at Claire in disbelief. Nico, as was his custom, bowed his head slightly in appreciation. I became aware, that when it came to Claire I felt neither competitiveness nor jealousy. With anyone else I would have berated myself for not being the first with the answer, but not with her.

With the cardinality and power set preliminaries out of the way Nico proceeded to the juicy parts. "I claim," he announced, "that the cardinality of the power set is always greater than the cardinality of the corresponding set."

At first glance this seemed totally obvious to me. We had just seen that for a set of cardinality n, the power set has cardinality 2^n, and $2^n > n$ for all nonnegative integers n. So it didn't seem to be as profound a result as Nico was making it out to be.

Nico had seen my head-shaking and nose-scrunching. "Ravi, the result is apparent for finite sets, but I'm saying that it even holds for infinite sets."

Ah, so! Now things were falling into place. This would mean, for example, that the cardinality of the power set of real numbers is greater than the cardinality of the reals. Here was the set that Claire and I had been speculating about. But how could one possibly get one's arms around the number of subsets of an infinite set? I tried to imagine the set of subsets of the real numbers and they rapidly spun outside what I could comprehend or even imagine. How, then, was Cantor able to show that the cardinality of the power set is greater than the cardinality of the corresponding set?

From my current perch in retrospective adulthood I consider Cantor's proof about power sets to be one of the flagship creations of the human race. We have thought nothing more elegant or powerful, only different. I reproduce his proof as Nico sketched it out that morning.

To make life easier I'll refer to the power set of A as P[A]. We wish to show that the cardinality of P[A] > cardinality of A. Since A is contained in its own power set there are only two possibilities: (1) the cardinality of P[A] is equal to the cardinality of A or (2) the cardinality of P[A] is greater than the cardinality of A.

We wish to prove that the first case is impossible and hence the second option must hold. Nico proceeded by assuming the converse of what we want to prove (as we had several times before by then) and finding a contradiction. So we'll assume that there is a one-to-one correspondence between A and P[A] (a one-to-one correspondence is the definition of equal cardinality) and show how that assumption leads to a contradiction and that therefore no such one-to-one correspondence could possibly exist. This can only mean that P[A] has a larger cardinality than A.

Let's say A = {c, *, a, ?, #, q, t. . .}. The trailing dots are meant to indicate that A is an infinite set. Our one-to-one correspondence would pair an element of A with an element of P[A]. Recall that elements of P[A] are just the subsets of A. Here's what a one-to-one mapping between A and P[A] might look like:

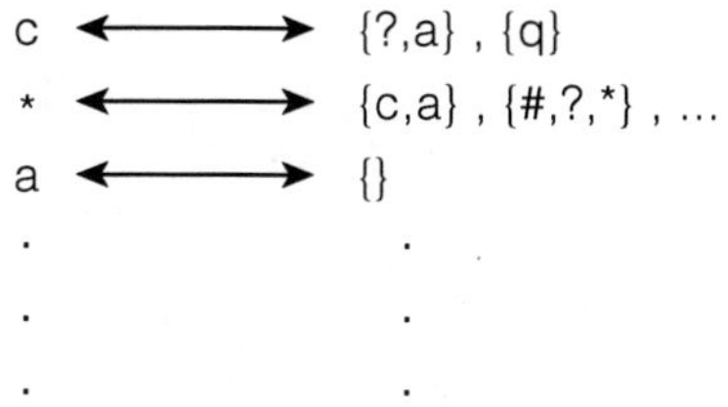

c ⟷ {?,a} , {q}
* ⟷ {c,a} , {#,?,*} , ...
a ⟷ {}
. .
. .
. .

This is only an example mapping but it allows us to observe some interesting patterns. Notice, for example, that the element c is mapped to the subsets {?, a}, {q}, and so c does not occur in the set it is mapped into. On the other hand * does occur in the P[A] entry because it is matched up with {c, a}, {#, ?, *}, ... Nico referred to elements that did not occur in their matched entry as "lonely" elements. Elements that did occur in their match entry were "happy" elements (that was not the adjective I would have chosen, but then Nico is an extrovert; to him togetherness equates with happiness). In our example c is lonely and * is happy. The element a is matched up with the empty set, so it is lonely.

Now we're at the point of pure genius:

Consider the set of all lonely elements. This set, let's call it L, is clearly a subset of A and so it is an element of P[A]. Therefore, in our one-to-one matching, it must have been paired up with some element of A. Let's say that L is paired with the element t.

Here's the question: Is t lonely or happy?

If t is happy it must be paired with a set that contains it. Since t is paired with L this is equivalent to saying that t belongs to L. But wait a minute! L is the set of all lonely elements. The happy t is prohibited from belonging to L. So t cannot be happy.

Conversely, assume t is lonely. Then t must belong to L since L is the set of all lonely elements. But if t belongs to L, then t is happy since t is matched up with L and the definition of "happiness" is for an element to be matched up with a set it belongs to.

Since t can be neither happy nor lonely we have a contradiction (it must be one or the other). This can only mean that our assumption of the existence of a one-to-one pairing between A and P[A] was false. Since we already know that the cardinality of A cannot be greater than the cardinality of P[A], the only possibility left to us is as follows:

cardinality of P[A] > cardinality of A.

Lovely.

This gets really interesting now. Start with the natural numbers N. One immediately gets another level of infinity by taking P[N]. But P[N] is itself a set. Therefore the power set of P[N] or P[P[N]] is a set of higher cardinality than P[N].

So now we have three levels of infinity: N, P[N], and P[P[N]]. But why stop here? We can do P[P[P[N]]] and P[P[P[P[N]]]] and so on forever. Each of these is an infinite set of ever-increasing cardinality! How sublimely amazing is that! Claire and I were looking to find a set of cardinality greater than c by looking at planes and cubes, but really we needed to be looking at power sets. We would have got not one, but an infinite number of ever-increasing levels of infinity.

"I have a question," said Adin. "Is the power set of natural numbers bigger than or smaller than the cardinality of the real numbers?" Basically Adin was asking for a comparison between c and the cardinality of P[N].

Nico liked the question. "What do you think?" he asked, confusing Adin a little, for he had spent not an inconsiderable amount of thought in formulating the question; he hadn't even got to thinking about the answer. But then Nico stared out of the window for a while and decided that the question was too complex for us to figure out on the fly. "Actually the two sets are exactly equal!"

Nico was saying that cardinality (P[N]) = c.

The proof he sketched relied on thinking in base 2. In base 2 you only use the digits 0 and 1. (In base 10 you use the digits 0 through 9.) Here is how you'd count in base 2: 0, 1, 10, 11, 100, 101, 110, 111, 1000, 1001 . . .

The advantage of thinking in base 2 is that it simplifies things. Instead of keeping track of various combinations of ten separate digits, we only worry about 0s and 1s. To demonstrate that cardinality (P[N]) = c, we may once again limit our attention to the set of reals between 0 and 1 instead of considering the entire continuum. (Again we rely upon functions that can take each element in (0, 1) to a point on the continuum and vice versa, so there is no need to think about the entire number line.)

Now the binary expansion (a cousin of the decimal expansion) of each point in (0, 1) can be represented by the set of all countably infinite sequences of 0s and 1s . Think of these as representing binary "decimals" between .000000 . . . and .111111 . . . . In this representation:

0.1 = 1/2,
0.01 = 1/4,
0.11 = 3/4, etc.

The power set of the natural numbers, P(N), can also be represented by the set of all countably infinite sequences of 0s and 1s. Each sequence represents a subset of N by interpreting a 0 in position n to mean that the number n is not in the subset, and a 1 in position n to mean that the number n *is* in the subset. For example, the set {1,3,5} corresponds to 0.10101 in binary; {} corresponds to 0, and the set of all naturals corresponds to 0.111111111111 . . . , which is another name for 1. Since the two sets have exactly the same representation for each of their elements we must conclude that they are essentially the same and therefore cardinality (P[N]) = c.

We had barely had time to catch our breath. In a few short weeks we had seen some amazing results:

1. Cardinality of the natural numbers ($\aleph_0$) = cardinality of rationals.
2. Cardinality of real numbers (c) > cardinality of natural numbers ($\aleph_0$).
3. For any set S, cardinality of P[S] > cardinality of S.
4. If S is an infinite set, P[S], P[P[S]] . . . are successively larger levels of infinity.
5. Therefore, there are infinitely many levels of infinity.
6. If N is the set of natural numbers and "c" is the cardinality of the real numbers, then P[N] = c.

We had indeed come very far. In fact we were now in a position to state the Continuum Problem, the most celebrated problem in all of mathematics. I will present it as I first heard it from Nico:

Continuum Problem: Is there a set whose cardinality is greater than the cardinality of Natural Numbers, but less than the cardinality of the Real Numbers?

A simple enough question. It's really asking if $\aleph_0$ and c behave like successive natural numbers. Just as there is no natural number between 0 and 1, is there no other infinite set between $\aleph_0$ and c?

Nico went on to observe that regardless of which set you might think would have a cardinality that lies between the natural numbers and the real numbers, it turns out to be at one end or the other, never in between. "We already saw that the rational numbers have a cardinality equal to $\aleph_0$. The irrationals turn out to have cardinality equal to c."

He went on to talk about a family of numbers he called algebraic that turned out to be $\aleph_0$ many, while the transcendental numbers turned out to have cardinality c. "In instance after instance there seems to be nothing in between. But at the same time there is no reason why there shouldn't be something in between!"

Once Cantor proved that there were infinitely many transfinite cardinals, he gave them names. $\aleph_0$ was followed by $\aleph_1$, then $\aleph_2$, $\aleph_3$. . . , and so on forever.

Given this series of cardinals Nico reformulated the Continuum Problem a little bit. "The question becomes: Is the cardinality of the continuum (or reals) c equal to the second cardinal $\aleph_1$?"

Continuum Problem (Reformulation): Is $c = \aleph_1$?

Nico stopped his pacing and stood by the poster of Cantor, which he'd affixed to the wall by the blackboard. He spoke in a low voice, looking at his shoes, not the photograph: "This man deeply believed that there were no cardinals between $\aleph_0$ and c, and that c was indeed equal to $\aleph_1$. His belief became known as the Continuum Hypothesis. It was a hypothesis because despite many years of intense effort Cantor was unable to provide a proof. Many times he thought he had a proof, only to have his hopes dashed. Early on in his investigations he even tried to prove that his hypothesis was false, that is, that there are cardinals between $\aleph_0$ and c, but those attempts didn't go anywhere either. As he grew older, Cantor developed an abiding faith that his hypothesis was true, but the lack of a proof continued to frustrate him 'til his death. Some say that the Continuum Problem contributed to the ultimate insanity of Georg Cantor."

• • •

Cantor's letter to Mittag-Leffler August 26, 1884

Dear Gösta,

Over the last many years you have been one of the only mathematicians who has taken an active interest in my findings about the infinite. Indeed, you were kind enough to shower high praise on my theorem that there are more real numbers than natural numbers. You said, perhaps carried away by the moment, that the diagonal argument was one of the most powerful and beautiful demonstrations you have ever seen.

Those words of yours have been a source of strength for me. I have taken solace in them in lonely moments when I have felt misunderstood or ignored by the majority of mainstream mathematicians. Please accept my heartfelt gratitude.

I am writing to you today to tell you about a very natural problem that has come up. It has been slippery and enormously difficult to solve. I call it the Continuum Problem.

In its basic form the Continuum Problem is simple to formulate. The diagonal argument shows that the cardinality of the continuum (real numbers) is greater than the cardinality of the set of natural numbers. Mathematically,

$c > \aleph_0$.

Now I ask, Is there a set whose cardinality is greater than $\aleph_0$ but less than c?

The question is deceptively innocent. You may think that its answer may well be achievable by elementary methods. Far from it. This has, by a wide margin, been the most difficult problem I have worked on. Its grip on me has been intense; I've eschewed food, sleep, and family in order to work on it.

Late yesterday night I saw my way out of the quandary. I have discovered that there is in fact no cardinality between $\aleph_0$ and c. The continuum is indeed the next level of infinity after the natural numbers (or the rationals). Without exaggeration, I feel sublimely rapturous this morning.

Later today I will carefully write out the proof and mail it to you.

Warm regards,

Georg

• • •

Cantor's letter to Mittag-Leffler October 20, 1884

Dear Gösta,

I am depressed and troubled. I told you that I had the solution to the Continuum Problem, but it turned out that there was an error in my approach. I have tried various resurrections and workarounds, but the problem has stubbornly refused to yield.

The quest for a solution now has a vice-like grip over me; it is impossible for me to think about anything or anyone else. I have not been out of the house for weeks and have not spoken to another human for days.

I regret my earlier communication giddily announcing a solution. The problem remains unresolved.

Sincerely,

Georg

. . .

Cantor's letter to Mittag-Leffler November 14, 1884

Dear Gösta,

I can live again! It gives me great pleasure to tell you that I have finally resolved the Continuum Problem. Much to my surprise (and contrary to my initial hypothesis) I have been able to show that there must be an infinite set whose cardinality lies between $\aleph_0$ and c.

My proof is unconventional, but this time I assure you it is correct. I'll be sending you the details later this week.

Warm regards,

Georg

. . .

Cantor's letter to Mittag-Leffler February 15, 1885

Dear Gösta,

The Continuum Problem is slowly but surely driving me insane. Every time I think I have solved it, I find a mistake in my solution. It pains me to tell you that the proof I referred to in my letter dated Nov. 14th was incorrect.

Confessing errors does not come easily to anyone, and confessing mathematical errors is doubly difficult for any mathematician worth his salt. Yet, I have found myself in this very situation not once, but twice.

I do not have a proof; I do not believe I will ever find one. I have come to believe that the Continuum Problem is extraordinarily difficult—it is almost mystical in some ways. It may never be resolved. For what it's worth I have developed a strong intuition that the continuum is indeed the first number class after $\aleph_0$, but this is a mere unproven hypothesis. In mathematics an unproven hypothesis is worth nothing.

Please accept my apologies for this entire correspondence.

Sincerely,

GEORG

• • •

On the way back home, acting on an impulse whose suddenness surprised even its own creator, I swung by the Law Library. The sullen document-custodian usually guarding the archives was not there. In his place was an old man, at least in his eighties, with thick reading glasses. He was bent over some older court transcripts, no doubt uncovering some history of his own. His jacket identified him as a university volunteer. The intelligence on his face gave me hope that he would be less rigid about the "one-transcript-per-week rule."

No such luck. "Says here that you checked out another transcript from the same case earlier today," he said. "Can't give you another one until next week." But his tone allowed room for discussion. I could tell he wanted to know what the big rush was. So I told him, emphasizing how much I wanted to know what happened to my beloved grandfather.

"This one time," he said.

• • •

MAY 14, 1919

Vijay Sahni vs. People of New Jersey

OFFICIAL COURT DOCUMENT

[Note from Court Reporter: All text below is authored by Judge Taylor. Judge Taylor ordered that the following be a part of the official case records.]

Below is a letter that I have written to Mr. Vijay Sahni. It contains my decision on the matter before us. I have asked that this letter be a part of the court records because it gives my considered response to the arguments put forward by Mr. Sahni over the last several days. Given the nature of these arguments, the philosophical questions he has raised, and the likely historical significance of this matter, I feel that it is important to include my thought process in reaching this decision.

May 14, 1917

Dear Mr. Sahni,

I have decided to recommend to the governor that your case proceed to trial. I am denying your request to drop the charges that have been brought against you. The Governor is scheduled to travel to Morisette in mid-June and I will be communicating my recommendation to him at that time.

I will do so with a heavy heart. Like you, I am not unaware of the likely outcome of your trial. The mathematical arguments you have made in our conversations can hardly be repeated in a court of law, and by your own admission, these arguments are your sole defense. A jail term for you certainly does not appeal to my sensibilities, but my job is to follow the law, and I'm afraid that the law is not on your side.

You may not agree, Vijay, but fairness and justice are American imperatives. We do not believe in unjustly condemning a man and neither do we allow any punishable offense to go by the wayside. The longing for justice flows in our blood.

Justice in this case was not clear-cut. Indeed I do believe that this is one of the most difficult judicial matters that I have wrestled with. The state blasphemy law was at odds with freedom of speech, and my task was to interpret how to apply these two competing laws. As you know, I had decided early on that if I found that you had made your remarks out of malice and forethought, then I would recommend a trial. On the other hand, if your remarks were merely knee-jerk responses driven by your cultural upbringing, or were caused by the heat of the moment, perhaps a response to an insult to your religion, then I would recommend an acquittal.

After talking to you it was quickly clear to me that your remarks were indeed made with forethought and contempt towards Christianity. You admitted this. Your defense was only that your contempt was justified.

Much as I was personally dismayed by your world view, I knew that I needed to listen and understand it. For, if your contempt were based on an open-minded evaluation without prejudice, then it would still be protected under free speech. Not everyone would have to agree with you, but your views would have to be tolerated. On the other hand, if I found your evaluation of Christianity to be deliberately closed-minded and willfully and maliciously intolerant, I would have to recommend a trial.

So I listened to you. I listened carefully. I found large portions of your arguments to be compelling and erudite. Indeed I am not lying when I say that you succeeded in changing the way I think about thinking. You succeeded in convincing me about the need for certainty in all spheres of human thought including religion. With Euclid's postulates you even succeeded in showing me a path to certainty. With the patience of a good teacher and the intelligence of a good thinker you took me a long way.

Yet in the end you fell short. The very intelligence that you used to open so many doors remained deliberately untapped when it came to religion. You were comfortable in applying your axiomatic approach when it came to geometry but were not willing to even consider it in matters of religion or spirituality. I could only ascribe your unwillingness to a prejudice against religion.

In a previous meeting, Mr. Sahni, you quoted from Descartes. I went back and read some more about him. Like you, he was passionate about certainty. Not the kind of certainty that says, "I am certain I will have supper tonight," but the certainty of ideas. The certainty that seeks to find bedrocks of truth that we can build on. Like you, Descartes was appalled to have accepted many false opinions as true and understood that without certainty there were only probabilities. And it is repellant to base the truly important things in life on probabilities. In fact he wrote:

> Throughout my writings I have made it clear that my method imitates that of the architect. When an architect wants to build a house which is stable on ground where there is a sandy topsoil over underlying rock, or clay, or some other firm base, he begins by digging out a set of trenches from which he removes the sand, and anything resting on or mixed in with the sand, so that he can lay his foundations on firm soil. In the same way, I began by taking everything that was doubtful and throwing it out, like sand. . . .

Now that is something I could hear you say. Like you, Descartes found a home in Euclid's *Elements*. He saw in it the model of thinking that he sought to apply to all knowledge. Like you, he based knowledge on a set of simple postulates that were clearly and undoubtedly true.

But this is where you part company with Descartes. For Descartes went on to use his axiomatic methods to actually prove the existence of God. He truly accepted the methods of Euclid and applied them everywhere, without prejudice.

Let me be clear: I am not punishing you for not being Descartes! But it is telling that a mathematician of his caliber decided to use the very same axiomatic methods that you hold in such high esteem to glorify Christianity, not to repudiate it. Why did he do this? But more importantly, why did you so obstinately refuse? Because you are prejudiced against religion in general, and Christianity in particular. That is the only conclusion that explains your behavior. Let me give you an example: I asked you to consider a simple axiom: "Everything must be created by something. It cannot come into being from nothing." I recall that you rejected this axiom completely. In fact, you said that the axioms of geometry are self-evident in a much more immediate way. This was the impetus that made us look at Euclid's axioms in the first place.

Well I've seen Euclid's axioms now, and I can truthfully say that they are as self-evident as the axiom that everything must be created by something. Euclid's fifth postulate sticks in my mind in particular—the one about lines meeting on the same side on which the interior angles are less than two right angles. We talked about that axiom and I challenged you on it. Because the language of the axiom was complicated, the underlying thought was not as apparent as the other four. You replied by asking if I believed the axiom, and I thought about it—if I draw two lines in space slanting towards each other, they will eventually have to meet—so I said yes I do believe the axiom, and we went on to study the actual propositions.

That was an example of an open, accepting mind on my part. Compare that to your rejection of the axiom I proposed. You did not even consider it! And this when the two axioms are equally self-evident. I am equally certain of both. But you choose to be blind to one.

It is this blind spot of yours that is forcing me to recommend that we proceed to trial. I wish this were not the case. You gave me the gift of geometry and it saddens me that things did not end differently. But justice must prevail.

Sincerely,

JOHN TAYLOR

• • •

So the Judge was using Bauji's own axiomatic method to find against him. It looked bad, but what gave me hope was that there were two more transcripts after this one. If this transcript contained John Taylor's parting words, it would have been the last one.

CHAPTER 6

THE NEXT MORNING CLAIRE CALLED and invited me over for lunch. I asked if I could come over right away and help her cook, and to my delight, she readily agreed. I took some lentils and spices and biked over to her place. Happily, we worked well together in the kitchen. Despite the tight space we didn't get in each other's way and without having a "you do this, I'll do that" conversation, we seemed to divide the tasks at hand pretty well. Later we feasted on Indian lentils over rice and fresh mint chutney from her herb garden. With satisfied palates and full tummies we sat and talked and played records and read to each other. Afterwards, I had a paper to crank out, so I parked myself at her table while she gardened. But I couldn't concentrate on the work at hand; my eyes kept drifting towards her. Today she was focused on protecting her new shoots from the unseasonable cold spell we were having. Her movements were precise and spare, and her single-mindedness was obvious. It was clear that she was completely in the moment and was fully occupied with what she was doing.

It occurred to me that Claire felt no ambiguity in what she was doing. She was quite certain of what needed to be done, and that's exactly what she did. The certainty that Adin was after was a different animal entirely. He wanted certainty at the absolute and objective level. Something that is timeless and context-free—universal and indubitable. Claire's gardening certainty was limited to her action and had the context of her backyard, experience, and aesthetics.

So there were different levels—or perhaps types—of certainty and some of them clearly existed. The question that Bauji and Adin were asking was if the grand prize of absolute certainty existed.

Later in the evening, when I told Claire about my encounter with Nico outside the cemetery, she made a similar point in a different way. "Ravi, Nico is so at ease with himself. He's either already asked and answered the kinds of questions that Adin is preoccupied with, or he has decided that they cannot be answered. Or maybe he has settled for a more personal type of certainty—something that is not absolutely true in the entire universe, but something that nevertheless allowed him to not let his mind drift into chaos after his wife's accident."

"Yes, but isn't the question of absolute certainty more important?" I asked.

"I guess it is. Intellectually, I think it is, but somehow I feel that there's something forced about the whole issue, and I can't put my finger on what it is. I mean, for me, Euclid's proofs were important, but endlessly worrying about the nature of truth seems a lot less immediate."

It didn't seem less immediate to me. Judge Taylor was ready to send Bauji to prison because they had conflicting notions of absolute certainty. "I wonder what's next in the transcripts," I said, more to myself than to her.

• • •

That night, armed with the date of Judge Taylor's letter and unable to sleep, I decided to take another look at *The Morisette Chronicle* microfiches. Perhaps some word of the case had made it to the newspapers. Not knowing what to look for the first time through, maybe I had flown right by something important.

It was after 2:00 a.m. when I got to the library. The sleepy undergraduate at the information desk expressed no surprise at a disheveled man requesting ancient copies of a rural New Jersey newspaper at that hour. Carol Stern would have wanted to know the whys and wherefores, but not this guy, at least not at 2:00 a.m. I followed him wordlessly to the scanner room, whose lock he opened on the fourth attempt.

In May 1919, judging by reports in *The Chronicle*, Morisette was occupied by news of the end of World War I. It was not that the town had developed a sudden interest in geopolitics; rather, it was the homecoming of the soldiers that had the town buzzing. The Selective Service Act was said to be rolled back, and a registration office in Newark was due to close operations in June.

Which is why Bauji's story had become less of a priority and had been relegated to the inside cover on page 2. I found the article in the May 17th edition, three days after Judge Taylor had declared his intent to let Bauji's case go to trial. It sat there innocuously sharing the page with news about a new steel plant near Princeton.

HINDOO'S BLASPHEMY CASE LIKELY TO PROCEED TO TRIAL

MORISETTE — Mr. Vijay Sahni's case is likely to be heard by a Morisette jury. We are told by reliable sources that Judge John Taylor has analyzed the facts of the case and has found no reason to dismiss the blasphemy charges leveled against Mr. Sahni. Court papers filed by Judge John Taylor are said to refute the prisoner's arguments.

Mr. Sahni, a Hindoo mathematician visiting Morisette to pursue mathematical learnings, was arrested on April 4th for making a speech judged by many locals to be contemptuous of Christianity and Christians. Sensitive to the need for upholding the laws of free speech, yet unwilling to let the Hindoo's blasphemy pass unnoticed, Governor Williams appointed Judge Taylor to analyze the facts of this unusual case and to provide a recommendation on how the state should proceed.

Over the past several weeks Judge Taylor and the court reporter, Mr. Hanks, have been seen entering and leaving the county prison where Mr. Sahni is housed. They have both been silent on the nature of their conversations with the prisoner. However, on this past May 14th, Judge Taylor is said to have decided that Mr. Sahni must stand trial for his inflammatory words and actions.

Many in Morisette have speculated that a jury trial will be beneficial to the county prosecutor. Sentiments against the prisoner have run high in this town and while the town has turned its attention to other matters, Mr. Sahni's name continues to evoke deep distress and indignation. This indignation will now have its day in court.

The next day's newspaper had a curious report that *The Chronicle's* staff didn't quite know what to make of.

LIBRARIAN REPORTS GEOMETRIC COMMUNICATION IN SAHNI CASE

MORISETTE — University Librarian Mr. Dwight Graham has reported that Judge Taylor has mailed Mr. Sahni a collection of geometry texts as well as a personal note concerning the upcoming blasphemy trial of The State of New Jersey vs. Vijay Sahni.

"The texts were on non-Euclidean geometry," said Mr. Graham, a statement verified by the library records. According to Mr. Graham, two weeks ago Judge Taylor spent several hours in the mathematics section of the library and finally selected works by the geometers Girolamo Saccheri, Janos Bolyai and Nikolay Ivanovich Lobachevsky. Yesterday, he was back at the library to renew the books and at that time he reportedly asked Mr. Graham to send the selected books to Mr. Sahni in the county jail. Mr. Graham maintains that the books were accompanied by a note from the Judge to the prisoner.

When asked what Judge Taylor wrote in the note, Mr. Graham told *The Chronicle*, "He asked the prisoner to closely read the books he was sending. He said that non-Euclidean geometry had an immediate bearing on the matters that they were discussing."

Legal experts and followers of this case tell the Chronicle that Mr. Graham's assertions are hard to refute, but are also difficult to understand. "What possible relationship could non-euclidean geometry have with the matters of faith that were likely discussed between Judge Taylor and the prisoner?" asks Mr. Seth Parker, a legal scholar from Newark who has been following the case from the beginning. "What is more likely is that the Judge sent over the books to enable Vijay Sahni to constructively pass the time in prison. He is, after all, a mathematician. Perhaps this is how he keeps himself entertained."

Judge Taylor could not be reached for comment.

It would turn out that Seth Parker was wrong. Bauji and Judge Taylor were on the cusp of learning about non-Euclidean geometries, and through their conversations, Adin, Claire, and I were about to get our first introductions to the subject as well. Belying its daunting name, non-Euclidean geometry would turn out to be one of the most touching episodes in the history of mankind's thirst to understand the universe we inhabit. We would see acts of great heroism and genius, but also failures of imagination and the bitter trampling of hopes and dreams.

It all began with the fifth postulate. Euclid, and just about every geometer for the next two thousand years after him, thought that the postulate was too complex. It read like a proposition that required proof, not a self-evident fact to be believed because it was entirely obvious. Euclid tried to prove it from the first four postulates but failed, so reluctantly, he included it with the other four.

• • •

EUCLID — JOURNAL ENTRY, DATE UNKNOWN

Between fitful bouts of sleeplessness I had a vivid dream last night: Scrolls of The Elements *were neatly laid out on a long rectangular table in a piazza that opened into a large garden. One by one, venerable scholars of geometry came into the piazza from a room at the far end. Each scholar picked up a scroll and walked into the garden. Strolling with the scrolls they pointed to and discussed the propositions of Book I of* The Elements. *These men were all skilled in geometry and had proved many difficult theorems, but even they were not immune to delighting at the logical structure of the book. One of them promised that my text would be translated into languages spoken in distant lands and that it would be read and revered by people for many centuries to come!*

This, I realize, is only a dream. What is more likely to happen is that the scrolls of The Elements *will be found in my room after I die and will be used for kindling that night. The copies I plan to provide to Tantalos and the others at the university will sit in their rooms unopened and will eventually be thrown away. I can see them now shaking their heads, remembering me with pity from time to time, "Old Euclid was really smitten with this idea of certainty in his old age. Poor fellow, after all that geometry, he lost his mind chasing an eccentric notion." Are they right? Am I mad to expect humans to achieve certain and absolute truth?*

I am not mad; I am succeeding.

I am succeeding in basing knowledge not on experience, but on logical thought. Setting modesty aside, I believe this is a revolutionary way to understand the world! I have discarded experience, for experience by itself can never provide certainty—different people can draw different conclusions from the same experience. Logic, on the other hand, is consistent and lasting. It is the only path to truth.

Only one problem remains. It is the annoying fifth postulate.

I have thought about it so often that I could recite it with nary a care about what the words mean: "If a straight line falling on two straight lines makes the interior angles on the same side less than two right angles, the two straight lines, if produced indefinitely, meet on that side on which are the angles less than the two right angles"

It is not that I believe the postulate is not true—no, I believe it completely. I know that if you draw lines in space that even slightly, ever so slightly, slant towards each other, they will eventually meet when extended far enough. It is an obvious property of the space we live in. Instead, it is the complexity of the postulate's thought that bothers me. A statement that is not completely and immediately apparent should not be a postulate. It should be a proposition, demonstrable from other postulates or common notions.

I have tried in every moment of wakefulness to prove the fifth postulate from the other four. I have begged and cajoled it, massaged it, even willed it to be demonstrated, yet it has resisted all my efforts. Sometimes I think that I almost have a proof, and then just as I am about to shout out in joy, I discover a flaw.

At times I am almost convinced that the information contained in this postulate is somehow independent of the other four. But then other times, when I am more confident of my powers, I think that there must be a way I could somehow succeed. The quest has driven me to exhaustion. I cannot sleep for more than an hour before I'm awake with the same thought, "Is there a way to demonstrate the fifth postulate?"

Today I have decided that my all-consuming quest must end. I cannot sustain this obsession any longer, or it will drive me insane.

Despite much temptation, I have avoided using the fifth postulate in the first 28 propositions of Book I. But now I am at a point where not using it means not making progress. Interestingly, assuming the fifth postulate immediately gives me all the power I need to complete Book I of The Elements. *Theorems of great simplicity and power fall quickly into place.*

For example, it immediately follows that a straight line falling on parallel straight lines makes the alternate angles equal to one another.

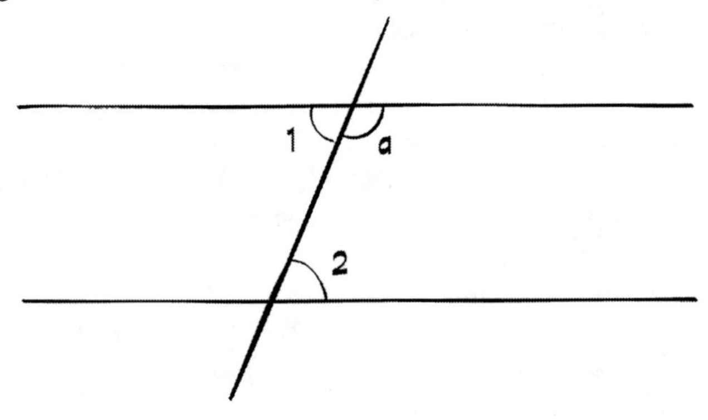

In other words, I want to show—using the fifth postulate—that Angle 1 = Angle 2.

The proof is really an immediate application of Postulate 5. Assume, for example, that angle 1 > angle 2.

2 right angles = angle 1 + angle a.
[Note from translator: 2 right angles = 180°.]

(The angles on a straight line always equal two right angles—I proved this in Proposition 13. It is, of course, a fact well known to even the students newest to geometry.)

2 right angles = angle 1+ angle a > angle 2 + angle a (since we assumed angle 1 > angle 2).

Now we use Postulate 5, which is built for exactly this situation. Since the interior angles are less than two right angles, the straight lines must meet when extended. But this is a contradiction, since the lines were assumed to be parallel—and parallel lines cannot meet. So angle 1 cannot be greater than angle 2.

An analogous argument establishes that angle 2 cannot be greater than angle 1. Since neither angle can be greater than the other, the two angles must be equal.

This is yet another example of the usefulness of the method of proof by contradiction: assume what you do not wish to prove and derive a contradiction. Since contradictions cannot exist (a thing cannot be and not be at the same time), you have established that the thing you did not want to prove is not true, thereby demonstrating what you wanted to prove in the first place.

The converse of this theorem is true as well. In other words, if a line subtends equal alternate angles on two lines, then the two lines must be parallel. To see this, use the method of contradiction again: assume the converse and use the fact that an exterior angle of a triangle, in which one of the sides is extended, is greater than the interior or opposite angles.

These two propositions—a parallel line must have equal alternate angles and conversely parallel alternate angles can only be subtended on parallel lines—unlock the keys to the kingdom, for everything else in Book I falls into place after their arrival. Immediately we get a proof that the sums of the angle of a triangle must equal two right angles.

For, given any triangle, draw a line parallel to its base. (I have a proof to show that given any line it is possible to draw a line parallel to it.)

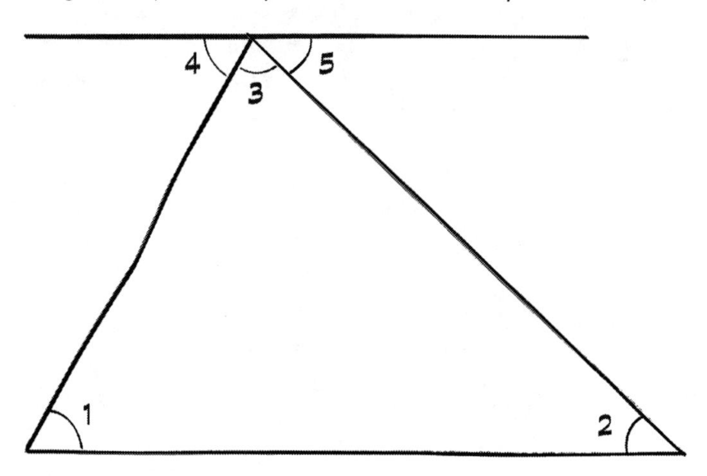

Now angle 1 = angle 4 and angle 2 = angle 5 (alternate angles). And, angle 5 + angle 4 + angle 3 = two right angles, for they are the angles, on a

straight line. Replacing equal angles, we have angle 1 + angle 2 + angle 3 = two right angles.

This result has been known for the last five centuries, but here at long last is a real demonstration. And it is an amazing result! It means that you could go anywhere in the universe and draw a triangle of any size and if you could somehow measure the angles of that triangle you would get two right angles.

Here is a certain truth, something we can rely upon. But even as I write this I am annoyingly reminded that it all rests on the fifth postulate. A postulate that is true, but irksome nonetheless.

If I were to magically awake from my grave a hundred years from now, my first question would be if someone has demonstrated the fifth postulate from the first four.

• • •

I was to find out that many generations of geometers after Euclid continued hunting for a demonstration—and they all failed repeatedly. Finally—and relatively recently—a new and revolutionary idea emerged: There may be other conceivable geometries in the universe that could be created by denying the truth of the fifth postulate! One would have thought that these strange geometries would run into trouble, that there would be a contradiction in them. But the suspected contradiction never emerged. These non-Euclidean geometries (so called because they denied Euclid's fifth postulate) seemed as consistent as Euclid's own geometry.

It is fair to say that the demonstration of non-Euclidean geometries is also one of the most significant moments in human thought. New philosophies arose to account for their existence, and they were to play a large part in modern physics as well. But Adin, Claire, and I didn't know any of this on the day we met at Tressider's for lunch to go over the next set of transcripts. In ways large and small, we were going to have our world views changed forever.

June 2, 1919

Vijay Sahni vs. People of New Jersey

Official Court Document

Judge Taylor: Did you read the books on non-Euclidean geometry I sent you?

Vijay Sahni: Yes, I did. They are fascinating—I have hardly been able to sleep, so fantastic are the results! And to think all these years I have ignored this subject as a mere curiosity. The few times I tried to study the subject I did not have the context I bring to it now. I always thought it was just a game with no ramifications to the wider stream of mathematics. But tell me, how did you come upon these books?

JT: As I have mentioned earlier, you have awakened in me a great interest in geometry. Your passion for the subject is infectious and it turns out that you are an inspirational teacher. After our discussions I, like hundreds before me, became smitten by the idea of proving Euclid's fifth postulate from the other four. You had shown me some demonstrations whose goal is to show a contradiction after assuming the converse of what must be proven, so I applied the same technique to the fifth postulate: I assumed its converse and sought a contradiction.

VS: This was before you read these books?

JT: Yes, it was. I had the same idea that seemed to have led to the development of the subject but I got stuck after assuming that the fifth postulate is false, and I didn't quite know what to do next.

VS: Nevertheless it is to your extreme credit that you came up with the basic insight that led to the creation of non-Euclidean geometry.

JT: Thank you. I find it remarkable that even though I have communicated my intent to decide against you, you seem to hold no ill will towards me. I was expecting you to be sulking, or even rude, yet here you are complimenting me.

VS: A good idea is a good idea, wherever it comes from. Anyway, so you got stuck and went to the library for inspiration?

JT: Precisely. And at the library I discovered Saccheri, then Bolyai and finally Lobachevsky. I read their work understanding little of the detail but much of the gist. However, what I understood was certainly enough to inspire me. I became anxious enough to feel a need to discuss the ideas I had learned about with someone. I went to the university to find someone who knew something about the subject but had no success. I could not find your host, who seems to be the only one who keeps up with current trends in mathematics. The others have barely heard of the subject and tend to think of non-Euclidean geometry as a theoretical exercise.

VS: As did I before I read these books. Why did you send these books to me?

JT: Because you are the only person who I knew would be interested, and I needed to discuss this with someone.

VS: So this has nothing to do with my case? Then why is Mr. Hanks writing a record of our conversation?

JT: Mr. Hanks is here because this conversation concerns a matter before the court. It is a continuation of the conversations we have had and it would be improper of me to talk to you without making a record of it. And as for the implications of this on your case, let's get to that after we more fully understand the nature of this new subject.

[Note from Court Recorder: Vijay Sahni sat lost in thought for a considerable time. The judge waited patiently for him to speak.]

VS: Where should I start? Maybe I should just sketch out the subject as I understand it?

JT: That would be fine.

VS: Reading these books you sent me, I came to realize that the question that I had dismissed as a mere irritant—that of proving Euclid's fifth postulate using the first four—has engaged some of the finest mathematical minds for over two thousand years. Proclus, the historian, commenting on attempted proofs to deduce the fifth postulate from the other four, noted that Ptolemy had produced a false "proof." He gave a good explanation of Ptolemy's error but then went on to give a false proof of his own! But in his meanderings Proclus did find a postulate that is equivalent to the fifth postulate, and this form of the postulate turned out to be useful in the development of non-Euclidean geometry. Here's how Proclus presented his version of the fifth.

> Given a line and a point not on the line, it is possible to draw exactly one line through the given point parallel to the line.

As sometimes happens in mathematics the reformulation of a problem or a proposition can open up new ways of thinking, illuminate a path where none seemed to exist before.

JT: Well, Proclus' formulation does somehow seem more intuitive to me, but I have learnt at my peril that we cannot take its equivalence to Euclid's fifth postulate at face value. How does one show that the two formulations are equivalent?

VS: Good question. The two postulates are equivalent because they contain exactly the same information. Each implies the other. By now you can guess how we demonstrate this.

JT: Assume one and try to prove the other, and then do the reverse?

VS: Exactly! In either case you are free to use the first four postulates, and hence the first 28 propositions in Book I, as givens. Recall that Euclid never used the fifth postulate in the first 28 propositions, so if you are given the first four postulates, you get the first 28 propositions as well.

Okay, let us assume Euclid's version first. So now you know that two lines with interior angles less than 180°, if produced indefinitely, meet. Now we must show Proclus' version: that given a line, call it L, and a point P not on the line, there is exactly one line parallel to L.

[Note from Court Reporter: Judge Taylor interrupted the prisoner in a loud excited voice.]

JT: I think I see how this follows! I find it amazing that with practice this step of conjuring up a possibility out of nothing seems to become intuitive. Before I began this study I would not even have known where to begin. But now it seems natural to begin with a construction.

(Note from Court Reporter: Judge Taylor drew as he spoke.)

JT: We can start by drawing a line through P that is perpendicular to L. Call this line m.

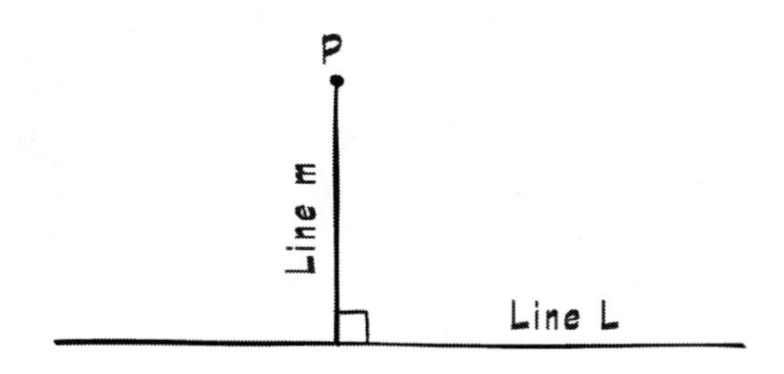

Euclid's eleventh proposition in Book I shows how to construct such a perpendicular. Now construct a line N perpendicular to line m.

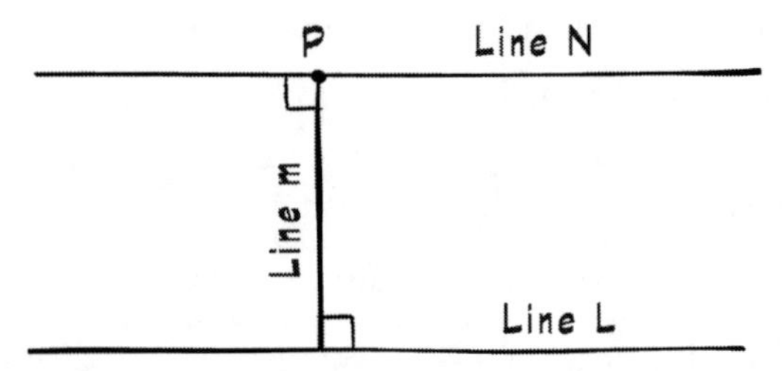

Now line L is parallel to line N because the alternate angles are equal; they are both 90°, and by a theorem of Euclid, equal alternate angles imply parallel lines. So there is at least one parallel line through P.

To show there is exactly one, just assume another parallel line through P. Call it line O. This line makes interior angles less than 180°—on one side of P, or the other—and since we have assumed Euclid's form of the parallel postulate, line O must meet with line L and hence cannot be parallel to it.

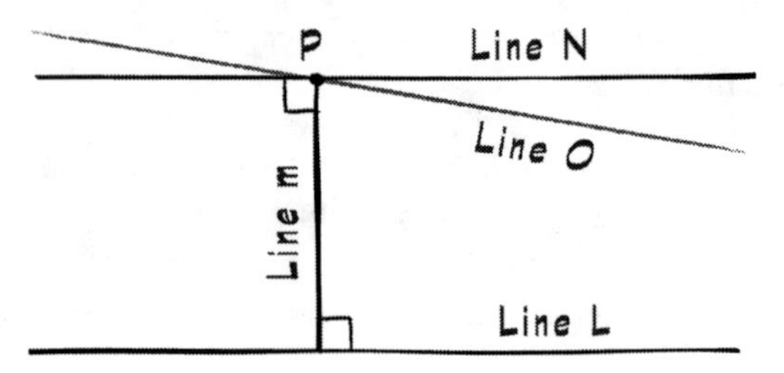

So no such line O can exist, and there can only be one line parallel to L that goes through P.

VS: Judge, that is great! Precise and to the point. You would make a good mathematician indeed. It is also possible to prove Euclid's form of the parallel postulate assuming Proclus' form and the original four postulates. So now you are assuming that there is only one parallel line through the point P and want to show that two lines with interior angles less than 180° will meet. The demonstration is simple and I need not spell it out for you. I will only say that it uses the path of contradiction: assume that what you are trying to prove is not true, and demonstrate that it cannot be.

JT: Yes, I can see how that would work.

VS: Good. Besides Proclus' there are other formulations of Euclid's fifth postulate:

1. If a straight line intersects one of two parallels, it will intersect the other.

2. Straight lines parallel to the same straight line are parallel to each other.

3. Two straight lines which intersect one another cannot be parallel to the same line.

And there is, of course, the one we just saw:

4. Given a line and a point not on the line, it is possible to draw exactly one line through the given point parallel to the line.

Any of these statements plus the other axioms imply Euclid's fifth, and vice versa. There are several other alternative formulations as well.

Many of these formulations came about as mathematicians sought ways to demonstrate the fifth from the first four. On many an occasion they would think that they had succeeded, only to discover that they had implicitly assumed an intuitive—but unproven—statement which would turn out to be equivalent the fifth postulate. One of the most systematic attempts at proving the fifth postulate was made by Girolamo Saccheri, an ordained Catholic priest, who ended up laying the foundations for non-Euclidean geometry. In my mind he is an unsung hero in the history of mathematics. I can't relate the story better than he does in his own words.

(Note from Court Reporter: The prisoner handed over a book to the judge, indicating the page he should turn to. The pages read by the judge are reproduced below.)

GIROLAMO SACCHERI — PERSONAL NOTES, MARCH 1722

My story is complicated but I am compelled, perhaps due to an exaggerated sense of my importance in the course of mathematical history, to record it.

Almost 30 years ago to this day, I was studying theology at the Jesuit College in Milan. I was going to the university, when I saw a slender man with a stick in his hand staring at the ground. There seemed to be purpose and intensity to his motionless staring, and so I went over to see what he was doing. He was looking at a triangle drawn in a semicircle.

"What are you thinking?" I asked him.

"I am thinking that a triangle must always subtend a right angle on a circle," he said. Then he looked at me and laughed at my confusion. But he laughed in a kind way and he invited me to sit beside him, patting down the spot he had chosen for me.

The man was Tommaso Ceva. He taught me mathematics. He was a great mathematician and a greater teacher. He was also a fine poet. I later came to see that he let his poetic sense guide his mathematical intuition.

Six weeks after meeting him I had mastered Book I of Euclid's Elements, *and my life had changed forever. After two decades of studying theology, a subject that flowed everywhere like a river in flood, it was a revelation to behold the rock-solid austerity of Euclid's geometry. I saw Book I like a sculpture by Michelangelo, but Tommaso said that it was like a work by Bach, starting simply with a few tone-setting definitions and modulating postulates, rising with each successive proposition where each movement had endless and surprising connections to the ones before it, and culminating in the great finale of the Pythagorean theorem.*

To me Elements *represented clean certainty, knowledge unburdened by the need for interpretation. Perfection.*

But there was one small imperfection: The fifth postulate. Not that it was not true. What reasonable man could doubt its truth? We intuitively know how space is and hence we know that lines in space must meet if their interior angles sum to less than 180°. The only open question is whether or not the fifth postulate is independent of the other four.

Postulates must be independent. There is no need for a postulate that can be derived from other postulates, for it should then just be a proposition. There is no need to assume what one can prove.

Euclid ultimately concluded that the fifth is an independent postulate, but other mathematicians through the centuries have tried to prove it from the other four. Many thought that they had succeeded, but really all they had done was to use an equivalent postulate without realizing it. Later mathematicians showed that the new property they had used was, in effect, the fifth postulate in different clothing.

This is how things have stood for two thousand years. But now I see a way—a way to remove Euclid's imperfection.

My way is inspired by the methods of Euclid himself: Assume that what you are trying to show is not true and derive a contradiction. Euclid did this repeatedly in both geometry and arithmetic. For example, to prove that there is no largest prime, he assumed there was such a creature and constructed a prime larger than the (assumed) largest prime. The contradiction showed that there could not be a largest prime. In general, the method involves making an assumption and deriving a result you know not to be true, and thereby concluding that the assumption itself was false. I have already used this method in a few simple discussions above.

I have determined that the way of contradiction is the best way to demonstrate the fifth postulate. I will assume that the fifth postulate is not true and try to derive a contradiction. As soon as I stumble on a contradiction, I will have demonstrated that the fifth postulate must be true. At that point we can safely call the fifth postulate a proposition. There would be no need to label such a complex thought a postulate, and Euclid would be forever free of imperfections.

An hour has gone by. I was rudely interrupted by a small fire in the kitchen caused by Giovanni, the cook, who is as excessive with his use of oil as he is with garlic. Apparently some oil from the fish sauce he was concocting dripped into the coals and started the commotion. The flames, no more than a few inches at their zenith, caused great panic in the quarters. Carmella, the maid, seemed near fainting, but I am thankful that things have now returned to normal.

As I was saying, I resolved to use the method of contradiction to prove the fifth postulate. This, by itself, is not a revolutionary idea, but the way I chose to apply it proved to be the key.

It turned out that, to proceed, it was convenient to re-state the fifth postulate. I focused my researches on a simple quadrilateral:

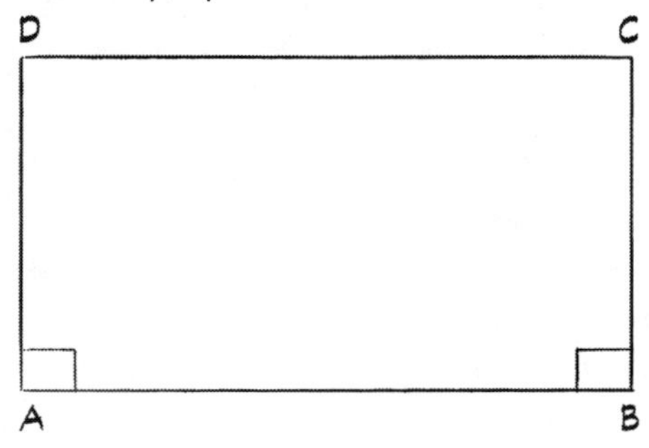

Very simple. Here is what is given to be true about this quadrilateral:

Angle A = 90°.

Angle B = 90°.

AD = BC.

What is not given:

Angle C = ?

Angle D = ?

The length of CD *relative to* AB.

I can almost see you scratching your head here! Why wouldn't Angle C and Angle D equal 90° and why wouldn't AB = CD?

The reason you expect this, dear reader, is because you intuitively accept the fifth postulate! If you assume the fifth postulate, then Angle C and Angle D would indeed equal 90° and AB would indeed equal CD.

But for my purposes I do not wish to assume the fifth postulate. I want to demonstrate it from the other four. So I may not assume that summit angles C and D are equal to 90°.

I started by making one important observation: Angle C = Angle D.

Again, this is obvious if you assume the fifth postulate, but not so obvious otherwise. To prove it we need to recall some simple theorems of Euclid concerning the congruence of triangles. Euclid proved these theorems without using the fifth postulate, which means that they are available for our use since we are assuming the first four postulates. These theorems were shown to me by Tommaso the first day he and I spoke:

> *Side Angle Side (SAS): If two triangles have, respectively, two sides equal, and the angle which they form is equal as well, then they are equal.*
>
> *Angle Side Angle (ASA): If two triangles have equal, respectively, two angles and the side which they contain, then the triangles are equal.*
>
> *Side Side Side (SSS): If two triangles have all three sides equal, then the triangles are equal.*

Notice that there is no AAA congruence rule. Two triangles may have all three angles equal, but their sides may be unequal. At any rate, we wish to prove that Angle C = Angle D in what I shall immodestly label the Saccheri Quadrilateral. Why do I wish to show this? Patience, dear reader, patience!

I will start by drawing the main diagonals in the quadrilateral. So, I will draw the line segments AC and BD. Here's what we get:

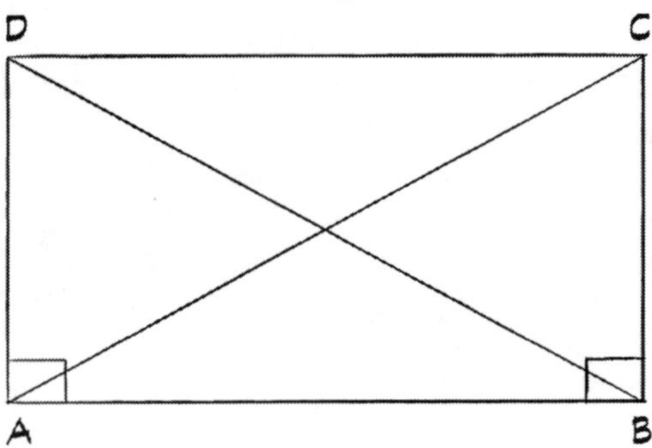

Now if we could somehow show that the triangle ACD = triangle BDC, we would automatically get angle C = angle D, for they would be corresponding angles in two congruent triangles.

But it is not readily apparent why triangle ACD should equal triangle BDC. None of Euclid's rules of congruence apply. All we can say is that AD = BC.

Fortunately, by the Side Angle Side (SAS) rule we can say that triangle ABC = triangle ABD. We know this because AD = BC (given property of the quadrilateral), AB = AB (of course!), and angle A = angle B (they are both given to be right angles). Since the two triangles are congruent, we may conclude that AC = BD.

This is exactly what we needed to show that triangle ACD is congruent to triangle BCD, for now all three sides of the two triangles are respectively equal and we may apply the Side Side Side (SSS) rule.

So we may conclude that angle C = angle D since congruent triangles have equal angles and sides, respectively.

This result was the key to my work, the key to removing imperfections from Euclid's masterpiece. For once I knew that angle C = angle D, I knew that there were only three possibilities:

1. *Both angle C and angle D are acute angles.*
2. *Both angle C and angle D are right angles.*
3. *Both angle C and angle D are obtuse angles.*

I will label these cases the hypothesis of the acute angle, right angle, and obtuse angle, respectively. I succeeded in proving that if there is even one case in which any one of these hypotheses is true, then it must be true in every case. Also notice that one of these hypotheses must be true, for there are no other choices!

Which brings me to the central idea of my work: if I could show that the hypothesis of the acute angle and the hypothesis of the obtuse angle lead to contradictions, then whatever is left—in this case the hypothesis of the right angle—must be true.

So what, you ask?

You would not ask if you knew that the hypothesis of the right angle must imply Euclid's fifth postulate. I was able to prove this in an unassailable and satisfying way.

Once again, if we eliminate the acute and obtuse hypotheses, we are left only with the right angle hypothesis. I have shown that this right angle hypothesis implies the fifth postulate, which is the very goal we have set out to reach. To be very clear, I will draw the implications we have created so far.

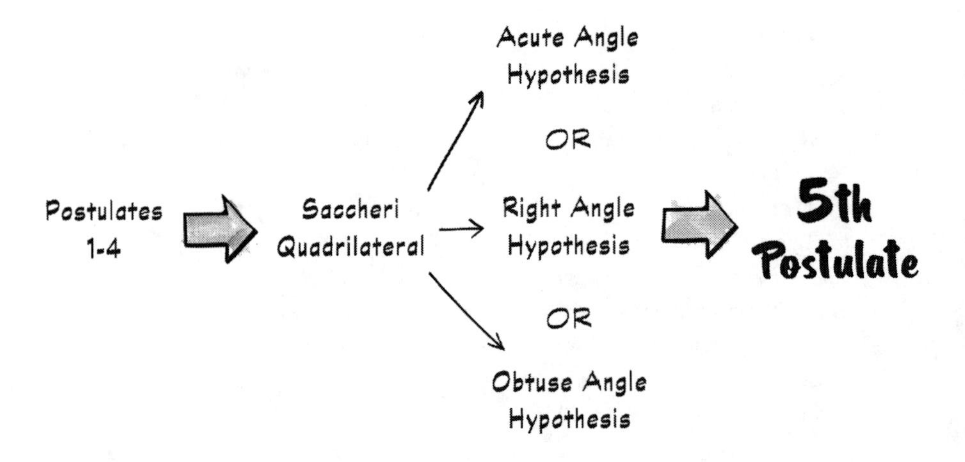

So it all hinges on deriving a contradiction for the acute case and the obtuse case.

I started with the obtuse case. After several pages of theorems that build upon each other (rather like Euclid's), I was able to show that the obtuse angle hypothesis (like the right angle hypothesis) implies the fifth postulate.

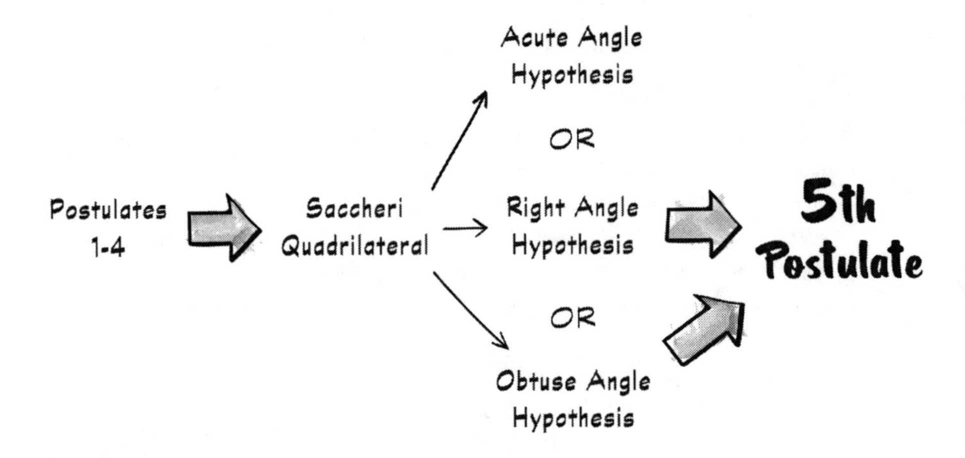

This is fantastic! Do you see why?

It is fantastic because we already know that the fifth postulate implies the right angle hypothesis, allowing us to make the implication arrow bidirectional.

So now we have:

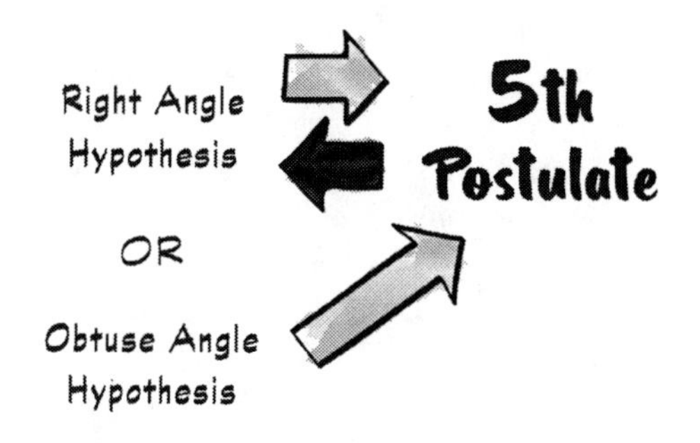

This is in effect saying that the obtuse angle hypothesis implies the fifth postulate, which in turn implies the right angle hypothesis. Eliminating the middle step, I have then "demonstrated" that the obtuse angle hypothesis implies the right angle hypothesis—but this cannot be! Either Angle C and Angle D are equal to 90° or they are greater than 90°, but they cannot be both at the same time. Therefore the obtuse case is false since it leads to a contradiction.

I have tried very hard to show that the acute angle hypothesis also somehow implies the fifth postulate. This would give me the other part of the contradiction that I have been seeking, but it has so far resisted my efforts.

I do not have a contradiction yet, but I feel that I am getting close.

• • •

November 1729

Seven years have gone by since I have written in my journal. In fact, I had even forgotten that I had started it. Seeking some scratch paper this morning I came across the above account, and I must confess it made me sad.

I was uncompromisingly hopeful then. I am more realistic now.

In the intervening seven years, I have derived theorem after theorem using the acute angle hypothesis. The theorems are strange and clearly untrue; nonetheless I have not yet derived the logical contradiction I have so patiently sought.

It is a perverse world, the one created by the acute angle hypothesis. There are triangles whose angles are less than 180°, lines that behave like curves, perpendiculars that behave impossibly—it is all repugnant to what we know about lines in space.

But repugnance does not a contradiction make—not a logical contradiction, anyway. I have spent hour after hour, night after night seeking the contradiction. Even in sleep I have dreamt about finding the refutation. My health has suffered, my loved ones have suffered, even my theological studies have been neglected—all to no avail. And really this is now enough.

I do not have the perfect refutation I have sought, but I will publish my work nevertheless, for there is so much progress here. I have refuted the obtuse case and I have shown the acute case giving rise to monstrous theorems that could not possibly be true. So, for all practical purposes the fifth postulate has been demonstrated and Euclid's one imperfection has been repaired.

I wish the repairs were stronger, though. I wish the refutation had been achieved on purely logical grounds, and not on the apparent impossibility of the results that stem from it.

Because it is still imperfect, I will not allow publication of this work until I die. And who knows? I might see the way to deriving a logical contradiction in the acute case in the years I have left.

[Note from Court Reporter: The judge read through the text with care, stopping at places to study the diagrams in great detail.]

JT: I can almost sense the agony Saccheri must have lived through. It is almost the despair of a believer who has begun to sense doubt, but cannot bring himself to recant.

VS: Perhaps the parallel is even more exact than you can imagine. Saccheri was by no means the only mathematician to undergo similar agony in his bid to prove the fifth postulate from the other four. There were others who were more willing to face up to the consequences of their doubt. Men such as János Bolyai and Nikolay Ivanovich Lobachevsky fearlessly went where their mathematics took them. I have selected some passages about these men that illustrate their passions and their mathematics. I would ask you to read through them.

[Note from Court Reporter: The prisoner handed over the material reproduced below for the judge to read.]

JT: I am eager to do so. I feel that we are on the verge of important things.

Letter from Farkas Bolyai to his son János December 1820

My dearest son, my pride and strength,

Sometimes life times things just so. Not three days ago I finished reading a book by the geometer Saccheri entitled *Euclid Freed of Every Flaw*. In this book, Saccheri attempts to prove Euclid's fifth postulate. It is clear that his work is one of much thought and monumental effort, yet on careful analysis it becomes apparent that, despite all his seeming progress, in the end Saccheri proved nothing.

He claims his contradiction to the hypothesis of the acute angle by saying that a particular property is "repugnant to the nature of the straight line." He throws up his hands and cries "Contradiction!" but really there is no contradiction at all; there are only theorems that appear strange to him. I do not fault him for this, because nearly a century after him we are not meaningfully closer to the answer he was seeking. As you know, my dear son, I too have spent countless hours trying to prove the fifth postulate from the other four.

Today, even as I was thinking about the meaning of Saccheri's lifework, and hence my own lifework, your letter came announcing that you, too, had decided to dedicate yourself to seeking a proof to the fifth postulate. Almost as soon as I finished reading your letter, I found myself with pen in hand writing to you with a plea from the depths of my soul.

You ought not to try the road of the parallels; I know the road to its end. I have passed through this bottomless night; every light and every joy of my life has been extinguished by it. I implore you, for God's sake, leave the lesson of the parallels in peace. . . . I had purposed to sacrifice myself to the truth; I would have been prepared to be a martyr if only I could have delivered to the human race a geometry cleansed of this blot. I have performed dreadful, enormous labors; I have accomplished far more than was accomplished up until now; but never have I found complete satisfaction. When I discovered that the bottom of this night cannot be reached from the earth, I turned back without solace, pitying myself and the entire human race.

I beg you, son, write poetry or plays, teach music or build homes, even grow apples or oranges if you like. In heaven's name, do anything except try to prove the fifth postulate. It is a problem that even the greatest mathematician of all, my friend Carl Friedrich Gauss, has been silent on. A problem that has silenced even the great Gauss is a problem of unearthly mystery and power; pursuing it can only lead to ruin. For God's sake, I beseech you, give it up. Fear it no less than sensual passions because it, too, may take all your time, deprive you of your health, peace of mind, and happiness in life.

If you never heed my advice on anything else, I will forgive you, but heed it now, dear son. Heed it now.

With abiding affection,

Your Father.

Letter from János Bolyai to his father, Farkas March 1823

My dear and good father,

Over two years ago you warned me to stay away from the theory of parallels. Much as I respect you, father, I must confess I have been disobeying you. For the last several months I have been deeply engaged in attempting to reach some conclusions about the absolute geometry of space, a geometry independent of the fifth postulate. You must not be too surprised, dear father, because I have learned my love for mathematics from you, and from you I have acquired a passion for attempting to resolve the question of the fifth postulate.

I am telling you this now because I have progress to report. As you had predicted, there were many frustrating nights and crushed hopes, but at long last I see the way.

I have now resolved to publish my work on parallels. . . . I have not yet completed the work, but the road that I have followed has made it almost certain that the goal will be attained, if that is at all possible. I have made such wonderful discoveries that I have been almost overwhelmed by them, and it would be the cause of continual regret were they lost before coming to fruition. When you see them, you too will recognize them. In the meantime I can say only this: **I have created a new world from nothing**. All that I have sent you 'til now is but a house of cards compared to a tower.

The Italian Bishop Saccheri was so close to being right. Using the hypothesis of the acute angle he discovered many theorems, but he kept looking for a contradiction, and that was his downfall. I have realized that there may be no contradiction at all; that not assuming the fifth postulate leads to a world that at first sight appears a little strange, but on deeper inspection is perfectly consistent.

Saccheri never found a contradiction, because I believe there is no contradiction to find. This new geometry seems as consistent to me as Euclid's geometry.

My approach has been to proceed without assuming the fifth postulate. Instead of assuming that there is exactly one parallel line that can be drawn through a point [which was Proclus' restatement of the fifth postulate], I assumed that there could be more than one line. This hypothesis, along with Euclid's first four postulates [which are not doubted by anyone], has given me a geometry that is not dependent on the fifth postulate. Theorem after theorem has followed with no hint of a contradiction. Now, admittedly, some of the theorems "feel" strange to our intuition, but this is only because our intuition is conditioned to believe the fifth postulate.

So, just as there was a Euclidean geometry, there is now a Bolyaian geometry! This is a victory for our family, father. Rejoice!

JÁNOS

JT: So János Bolyai seems to have made the jump that Saccheri could not.

VS: Indeed. He must have been a fiery young mathematician determined to make a mark for himself. He had the courage of his convictions and founded an entirely new subject. Equally inspiring is the story of the Russian, Lobachevsky, who worked independently from Bolyai, but came to similar conclusions. I will ask you to read his writings next. They will give you a flavor of some of the inner workings of non-Euclidean geometry.

[Note from Court Reporter: The prisoner handed over the material reproduced below for the judge to read.]

NIKOLAY IVANOVICH LOBACHEVSKY AUGUST 1855, KAZAN, RUSSIA

I am blind and sick now and I know I am dying. My dearest eldest son has already died, I am in debt, my marriage has failed, and my career is over.

But these are minor irritations compared to my greatest regret: the fact is that I have made the greatest mathematical discovery in two thousand years and have received no recognition for it.

The government did give me an award last month. The award was for a little machine I designed to process wool. I revolutionize mathematics and philosophy and they give me an award for processing wool!

More than 35 years ago I discovered that not assuming Euclid's fifth postulate gives rise to a new geometry that appears to be as logically consistent as Euclid's geometry. I am convinced that this is not just a consistent geometry interesting only to theoretical academicians, but it is the geometry of space. I have faith that space itself is non-Euclidean (although I cannot yet prove this).

But no one has understood the importance of this discovery. For over three decades I have tried to get people to read this work, I have translated it into French, German, and recently again to French, but it has not mattered. Nobody cares.

Not that I am unique in this fate. Saccheri, an Italian bishop, laid the foundations for non-Euclidean geometry but his work was not published until after he died and still no one quite understands what he did. Save for a small error when he let his desire to find a contradiction in the Hypothesis of the Acute Angle cloud his judgment, he let logic take him to theorem after theorem, each one as consistent as the last.

I've been recently told that a Hungarian mathematician by the name of Bolyai has made discoveries similar to mine. But he, too, is living in obscurity, when he should be recognized and congratulated.

Only the great Gauss has understood and praised my work. But that is little solace for someone who has made discoveries as revolutionary as Copernicus' discovery that the earth revolves around the sun and not vice versa.

Optimist that I am, I want to put a small flavor of my work in this diary of mine. My hope is that someday, somewhere, someone will look upon the words below and say, "That Lobachevsky really changed the way I think about the universe." Nobody looked at the mathematical articles, but perhaps someone will look at the diary.

The historian Proclus' formulation of Euclid's fifth postulate says, "Given a line and a point not on the line, it is possible to draw exactly one line through the given point parallel to the line." My idea was to deny the truth of this axiom. Instead of one parallel line through the point, I would assume two.

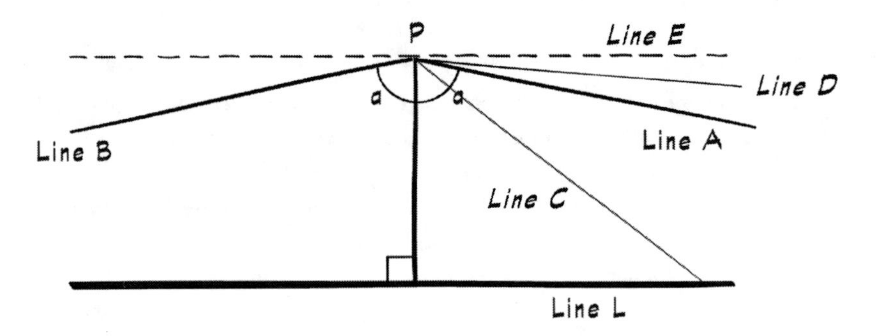

The picture looks complicated but really it isn't. Given a line L and point P, we draw a perpendicular between the line and the point. Now in Euclidean geometry there is exactly one line parallel to line L and that is line E. Line E makes an angle of 90° with the perpendiculars. By trying to draw more than one parallel line through P, we are assuming that there are lines, such as line D, that make an angle less than 90° with the perpendicular, but nevertheless never "cut" line L. On the other hand, there are lines such as line C that do cut line L. The cutting lines and the noncutting line must have a boundary line. This boundary line, which I've named line A in the picture, is the first line that does not cut line L. It makes an angle a with the perpendicular. Any line that makes an angle less than angle a will cut line L, and any line

that makes an angle greater than angle A will not cut line L. A similar logic applies on the left side, where line B is the boundary line.

I designate the boundary lines, line A and line B, to be parallel to line L. I also designate angle a to be the angle of parallelism.

I realize it is a strange construction. In the drawing, lines A and B certainly look like they will cut line L. Indeed, even line D looks like it will eventually cut line L. But we are investigating what happens when we make assumptions that go against our practiced intuition of the fifth postulate.

At first I thought that I would do what Saccheri tried to do: find a contradiction. But it slowly dawned on me that there would never be a contradiction. Just as Euclid deduced hundreds of geometric theorems in The Elements *without deriving a contradiction, I found that disciplined deduction took me to a new geometry, not to a contradiction.*

The first few theorems after the definition of parallel lines were quick and satisfying. I proved them over the course of one afternoon on a hot summer day some 35 years ago. I write them here without demonstration, but you must know that I did achieve a rigorous proof, a proof independent of Euclid's fifth postulate, for each and every one of them:

1. *A straight line maintains the characteristic of parallelism at all points, meaning that if line A is parallel to line L at a point P (i.e., the boundary line between the cutting and noncutting lines through P), then it is also parallel to line L at all other points.*

2. *Two lines are mutually parallel; i.e., if line M is parallel to line N, then line N is parallel to line M.*

3. *In a triangle the sum of three angles cannot be more than 180° (but as I will shortly demonstrate, it can be less than 180°).*

4. *If in any triangle the sum of the three angles is equal to 180°, so is this also the case for every other triangle.*

5. *From a given point we can always draw a straight line that shall make, with a given straight line, an angle as small as we choose.*

*And now I will prove a result that brings to light the incompleteness of the Euclidean system. I will show that, in general, the sum of the angles of a triangle is **less** than 180°.*

I will use Theorems 3 and 5 in the demonstration.

Pictorially:

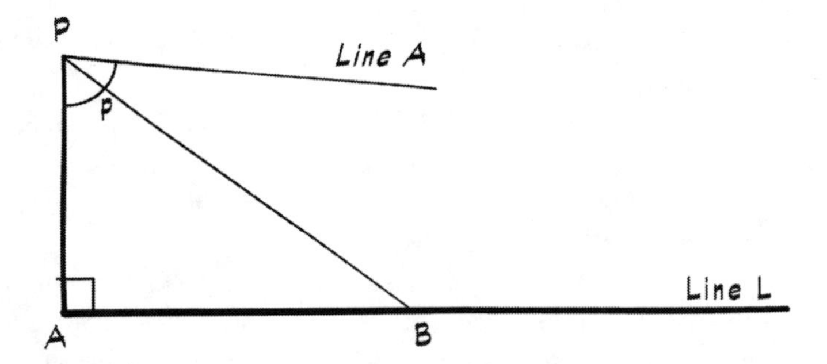

In the picture angle p is the angle of parallelism (less than 90°) and line A is parallel to line L. I claim that the sum of the angles in triangle PAB is less than 180°. We begin by drawing PC such that AC > AB. I will also name all the angles in the picture so that it will be easier to work with them.

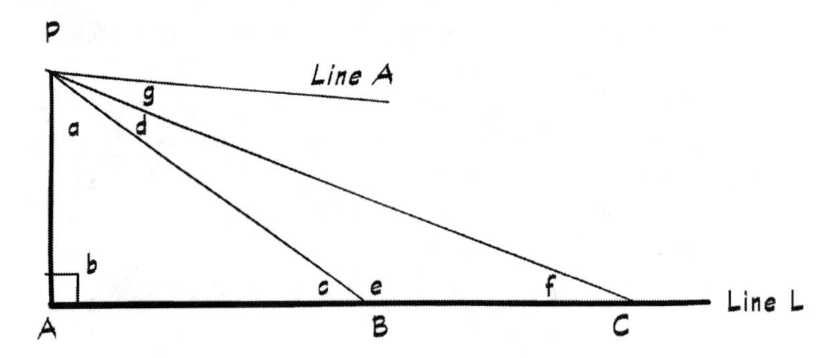

We want to show that if the angle of parallelism is less than 90° then angle a + angle b + angle c < 180°.
A few observations about the angles:

1. *Angle b is given to be 90°.*

2. *Angle c + Angle e = 180°.*

3. *Angle a + Angle d + Angle g = Angle p, the angle of parallelism.*

In triangle PAB, let's assume that the sum of the angles is 180° – X. Further, assume in triangle PBC that the sum of the angles is 180° – Y.

Note that Theorem 3 above guarantees that X and Y are nonnegative quantities. (Theorem 3 stated that in a triangle the sum of three angles cannot be more than 180°.)

We want to show that if angle $p < 90°$, then X is a positive (greater than 0) quantity, thereby making the sum of the angles in triangle PAB less than 180°.

Now consider the big triangle PAC: the sums of the angles of this triangle are all the angles in the triangles PAB and PBC except for angle c and angle e. Therefore, the sum of the angles in triangle PAC = (180° – X) + (180° – Y) – (angle c + angle e).

Since angle c + angle e = 180°, the sum of the angles in triangle PAC = (180° – X) + (180° – Y) – (180°) = 180° – X – Y.

Looking at the angles in triangle PAC in another, way we have the sum of the angles in triangle PAC = angle a + angle d + angle f + angle b = angle a + angle d + angle f + 90° = 180° – X – Y.

Therefore, angle a + angle d + angle f = 90° – X – Y.

Recall that angle a + angle d + angle g = angle p, where angle p is the angle of parallelism. Therefore, angle p – angle g + angle f = 90° – X – Y. Rearranging terms, we get: 90° – angle p = X + Y + (angle f – angle g).

Now, by drawing C far enough along line L, we can make angle f and angle g as small as we want. (This follows from Theorem 5 above, which stated that, "From a given point we can always draw a straight line that shall make, with a given straight line, an angle as small as we choose.") Therefore, the quantity angle f – angle g may be assumed to be zero. Hence 90° – angle p = X + Y. Since neither X nor Y is a negative quantity, at least one of them must be nonzero.

Assume X is zero. This means that Y is a nonzero positive.

If X = 0, then the sum of the angles of the triangle PAB, which we assumed to be 180° – X, is in fact exactly equal to 180°. Now, I have previously told you about a result that states, "If in any triangle the sum of the three angles is equal to 180°, so is this also the case for every other triangle."

This is implying that the sum of the angles of triangle PBC must also be 180°. But we had earlier seen that this sum was 180° – Y, where Y was a nonzero positive quantity. Therefore, X cannot be zero.

Notice that if the angle of parallelism, angle p, was equal to 90°, we would have: 90° – angle p = X + Y or 0 = X + Y, implying X = 0 and Y = 0. This would have shown that the sum of the angles of a triangle is exactly 180°, which is precisely what Euclid claimed.

But if the angle of parallelism, angle P, is less than 90°, the sum of the angles of a triangle would be less than 180°. This shows that Euclid's geometry is merely one specific case of the more general Lobachevskian geometry.

I hope this gives you a feel for the type of results that follow if you assume that Euclid's fifth postulate is false. There were, of course, many other results. Together they make a system that I believe is as consistent as Euclid's, but more than that I sometimes believe that it is also more real than Euclid's. I have a strong sense that my geometry is, in fact, the geometry of space . . . , that two light rays parallel—in my sense of the word parallel—extend forever without meeting, even though they slant towards one another at their origin. I will die with that image in my eyes.

(*Note from Court Reporter: In the silence that followed, Judge Taylor and the prisoner seemed preoccupied with their own thoughts. Each referred to the above materials and each drew out several diagrams in their respective notebooks. Judge Taylor was the first to speak.*)

JT: It's late and I am extremely tired. I cannot fully trust what my mind is telling me. Let us reconvene in the morning and continue our discussions.

VS: I agree. I need some time to further internalize all of this as well. I too will be fresher in the morning. Good night then?

JT: Good night.

• • •

Fortunately I didn't have to wait an entire week to get the transcript of the following morning's session. Mr. Hanks, or someone after him, had clipped the two records together, and hence they had been filed as a single document. Thank God, for I don't think I could have waited a single day for whatever was to come next.

Just before turning the page to the next morning's conversation I had the strange sense of being with Bauji in his cell in Morisette. I pictured that the Judge had just left after the conversation above and that Bauji and I were sitting on the floor, our backs resting against the wall, with a heap of books on non-Euclidean geometry spread out in front of us.

"What does this mean Bauji?" I would have asked him.

"I'm not sure yet Ravi," I could hear him say. "I'm not sure if this is some fantastic mathematics or if it really describes the universe we live in."

"But Bauji! These geometries are absurd. They are too strange. I keep thinking a contradiction will show up in them at any second."

I saw Bauji shake his head with his characteristic slowness. "There will be no contradiction," he said. "This geometry is logically sound. I am sure of it."

I wondered how he could be so sure.

• • •

June 3, 1919

Vijay Sahni vs. People of New Jersey

Official Court Document

Judge Taylor: Good morning Vijay. Did you sleep well?

Vijay Sahni: No Judge, I did not. I was quite troubled, as a matter of fact. I am not sure what to make of this new geometry. And I'm not sure what it implies for the truth of Euclid's geometry.

JT: Why? Has it shaken your confidence in your idea of certainty?

[Note from Court Reporter: Prisoner muttered unintelligibly and was asked by Mr. Hanks to repeat his statement.]

VS: I said I don't know.

JT: You don't know? You mean everything you have told me could be wrong?

[Note from Court Reporter: Prisoner did not say anything.]

JT: Well? Could it be?

VS: Yes. It could. That is what you were waiting to hear, is it not? Now you can feel better about your decision not to free me.

JT: No, no. That is not what I am after. I'm after understanding, understanding about what all of this means. Tell me, why do you feel that non-Euclidean geometry casts a shadow over your axiomatic method? I'm not sure I understand.

VS: I've always thought of the axioms as self-evident, things that had to be true, things that just could not be any other way. Assuming the negation of the fifth postulate should have led to an immediate contradiction. Instead, it leads to an alternative geometry—and the geometry appears to be consistent. It undermines my belief that the fifth postulate is necessarily true.

JT: No contradiction has been found yet—this doesn't mean it will not be found. Who knows? Maybe Bolyai and Lobachevsky were premature in their conclusions. Their geometry looks consistent, but it has not been proven to be so. When I say consistent, I mean contradiction-free.

VS: There is no proof that Euclid's geometry is consistent either.

JT: But we know Euclidean geometry is consistent!

VS: No, we do not. We believe it is consistent. In other words we believe that it has no contradictions, but we cannot prove that.

[Pause]

JT: I see. Intuitively that makes sense. For how could you prove the consistency of any geometry? I mean, you'll have to show there can be no contradiction no matter what you do.

VS: It is a tall order. But this may surprise you: I can demonstrate that if Euclidean geometry is consistent, then non-Euclidean geometry is consistent as well.

JT: Really? That could be extremely important. And I see why you are taking these geometries so seriously.

VS: Yes. It would show that from a logical standpoint Lobachevsky is as secure as Euclid.

JT: One surprise follows another these days, Mr. Sahni. I find it incomprehensible that a geometry that assumes that a line can have more than one parallel passing through a point is as valid as Euclid's geometry. How do you prove such a fact—or is it too difficult to elaborate upon?

VS: On the contrary, it is quite simple. The idea is to build a model of non-Euclidean geometry within Euclidean geometry. It involves not being wedded to what things mean in a deductive system.

JT: What?

VS: Let me give an example of a deductive system. This system has only one axiom: If a collection has a property, then all instances of that collection have the same property. That's the axiom. So now let us apply this axiom. Say we are given two facts: (1) All men have red eyes and (2) Judge Taylor is a man. Now we can apply the axiom to conclude that Judge Taylor has red eyes.

JT: Sure. So what?

VS: I'm not finished. Now notice that the deduction has nothing to do with the definition of red eyes, or what it means to be a man, or who Judge Taylor is. The logic works if instead of red eyes we used brown eyes or blue eyes or even bushy tails. The meaning of the words does not matter to the structure of the argument, as long as the axiom is satisfied.

JT: I agree. How does this apply to showing that non-Euclidean geometry is as consistent as Euclidean geometry?

VS: Euclid's geometry, like the example I just gave, is a deductive system. It uses logic on objects that obey the laws as laid out in the axioms. The precise definition of the object does not matter. So, for example, what the definition of a line is does not matter as long as the line meets the requirements of the axioms.

JT: Sure. I see that.

VS: Now what I will do is briefly outline a model where we can give a different meaning to the terms we used in the Euclidean axioms, terms such as plane, line, and point. Essentially we would merely rename certain objects in Euclidean geometry in such a way that a non-Euclidean geometry arises. This geometry must be as consistent as the original Euclidean geometry because it arises from a mere renaming of Euclidean objects. If the model has a contradiction, then we could get the same contradiction in Euclidean geometry merely by renaming the model objects to their original Euclidean names. And vice versa, if there is a contradiction in Euclidean geometry, we could get the same contradiction to appear in the model. But because the model itself would be non-Euclidean, we would have shown the logical equivalence of Euclidean and non-Euclidean geometries.

JT: So the trick is to create a non-Euclidean model by renaming some Euclidean objects.

VS: Precisely. And this is not difficult to do. Let me sketch it out. First let's just draw a simple circle. In my model I'll define a "plane" to be all points that lie strictly inside the circle. In other words I am postulating that the circle itself is not part of my "plane" but every point inside the circle is. It is not a plane in the traditional, Euclidean sense, but in our model, this is what we will agree to name a plane.

JT: Okay. I still don't see what that gets us, but keep going.

VS: You'll see in a few minutes. Next we need to describe what a line is in this model.

JT: Well, wouldn't it be natural to call any chord inside the circle a "line" in our model?

VS: Exactly! What we call chords in ordinary geometry will be the lines of our model. Again it is important to note that the end points are not part of what we now call a line. Let me draw a picture here.

[Note from Court Reporter: Prisoner drew figure reproduced below.]

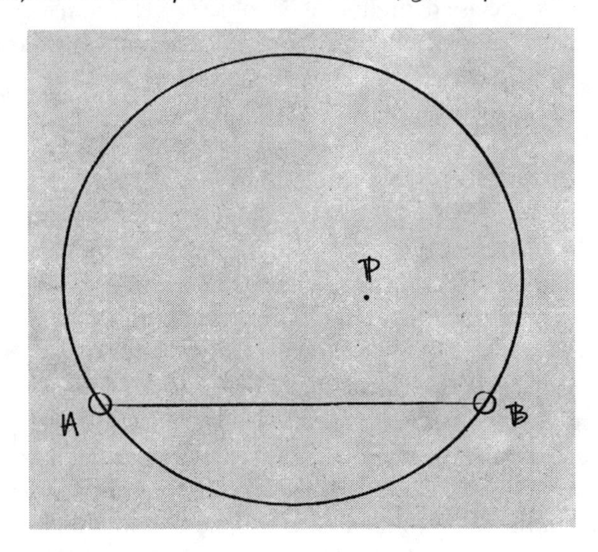

VS: In this picture, P is a point in my plane and the line AB is a line in the plane with the proviso that the points A and B themselves are not considered to be inside the plane.

JT: Is that why you've drawn little circles around A and B?

VS: Precisely. If you think about it, the model is really straightforward. The definition of a point does not change but for the fact that points must lie within the circle; a line is just your ordinary line but restricted to within the circle and designated by the end points which are not part of the model.

JT: But how can we be sure that this new meaning we have given to terms such as line allows us to make sense of the axioms?

VS: Ah, Judge, that is the beauty of it! It turns out that the axioms do make sense in our model. Take the first postulate. Is it possible to draw a line between any two points on our plane?

JT: Yes, because if the points are within the circle there is no problem drawing a chord between them.

VS: Exactly. Postulates 2 and 3 require us to modify our notion of distance. Recall that Postulate 2 required us to extend straight lines indefinitely and Postulate 3 required that we be able to draw circles of any radius at any point.

JT: We have a problem here. We're bounded by the circle. Our plane does not extend forever.

VS: You're right. But there is a way around the difficulty. We could redefine what distance means in this plane, so that given two points C and D their distance approaches infinity as D approaches the circle. There are some technical details to sort out in that the "distance" is required to be invariant in any non-Euclidean displacement, for displacement should leave distances invariant. But these are technicalities that I will ask you to ignore as long as you see the gist of the argument.

JT: Yes, I think I do see the gist. Anyway, moving on, the fourth postulate about all right angles being equal seems to go through without difficulty. Which brings us to the fifth.

VS: *[laughing]* Ah yes! The fifth. Things always get interesting with the fifth. Let me ask you, Is it possible to draw more than one parallel line through a point?

JT: I can't answer that unless we define what it means to be parallel in this model.

VS: Judge you have a fine mathematical mind. You didn't assume that the old definition of parallelism will go through in our model. But let me challenge you, What do you think a natural definition of parallelism might be in our model?

JT: It would seem to me that we should say that lines that do not intersect are parallel.

VS: Indeed. In fact, we will have parallel lines precisely as Lobachevsky defined them.

Watch:

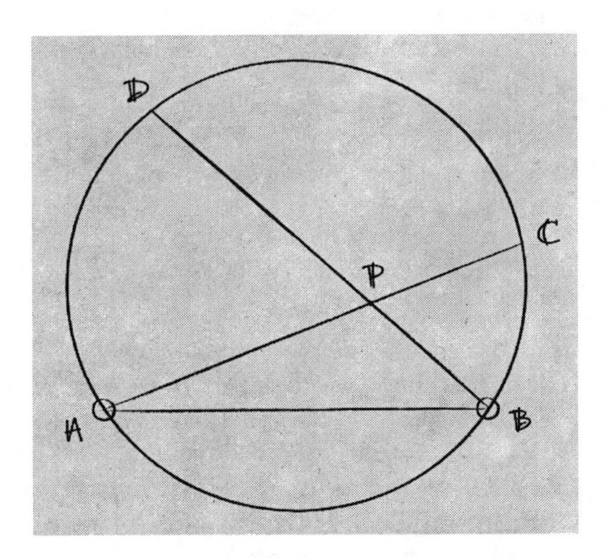

In this picture BD is parallel to AB because they do not intersect each other.

JT: Wait a minute. They do intersect at B.

VS: Yes, but the point B is outside the plane, remember?

JT: Yes, of course. So then AC is parallel to AB in the same way.

VS: Yes. But notice that we have two lines parallel to AB through a single point P.

JT: Which is a Lobachevskian geometry!

VS: Yes! But remember that this non-Euclidean, Lobachevskian geometry is a model within Euclidean geometry. We built it by renaming Euclidean objects and redefining the relationships between them. Therefore the theorems of this non-Euclidean geometry are merely different ways of phrasing certain theorems in Euclidean geometry. A contradiction in any one of these geometries must exist in the other.

JT: Hallelujah! I understand the argument. Yes, if there ever was a contradiction in non-Euclidean geometry, you could get the exact same contradiction in Euclidean geometry.

VS: Indeed. There are a few technicalities to sort out. In particular, we glossed over exactly how we would change the traditional notion of distance in order to leave it invariant under non-Euclidean displacements; but what we have discussed sketches out the essence of the matter.

JT: I see it but I do not believe it. You have shown that Bolyai's and Lobachevsky's geometry is as logically sound as Euclid's. You have also shown that the fifth postulate is independent of the first four, else assuming the negation of the fifth postulate would give one an immediate contradiction, which at long last answers the question that has plagued geometers through history. This model has been a truly powerful tool!

VS: Yes. The model argument is elegant but the conclusions are worrisome. Non-Euclidean geometry is a strange construct, yet it equals Euclidean geometry in logical soundness.

JT: Well, at least we know that non-Euclidean geometry is not really true in space. I mean, it is only true logically, in our minds. So, in that sense, your faith in the sanctity of the axiomatic method is still valid.

VS: But, Judge, I am not even sure that Euclidean is the geometry of space anymore.

JT: Of course it is! Are you making a fool of me?

VS: No, I am not. I wish I were.

JT: How can you say such a thing?

VS: Judge, perhaps you missed the passage in Lobachevsky's book, but he mentions that his geometry could, in fact, be the geometry of space.

JT: But you know it's not. We know that the fifth postulate is true in space. Are you telling me it's not? Are you telling me that a triangle has angles that sum to more than 180°.

VS: Judge, I don't really know. But I think it is possible.

JT: This is stupid. Let's get a protractor and actually measure the angles of a triangle. If they sum to 180°, we'd know Euclid was right about space.

VS: It's not that simple. Lobachevsky's geometry predicts that for small triangles, the sum of the angles is very slightly less than 180°. The defect is not measurable. It only becomes apparent for very large triangles. Triangles in space.

JT: I see. But I still think the idea is insane. If our basic intuition of Euclid's postulates can be wrong, then anything can be wrong. What is there to believe in?

VS: I don't know, Judge. I don't know.

• • •

Adin fretted while reading the transcript. His legs moved restlessly and without pause. From time to time he shook his head in disagreement. "There has to be only one parallel line you can draw through a given point. These other geometries don't make any sense."

"You can't change the fact that they exist," countered Claire. "And in their own way they are beautiful." But Adin shook his head again. How could something be beautiful when it didn't make sense?

The transcript had held us rapt. Our burritos sat uneaten, and we didn't notice Nico, with his dog Freud in tow, until they were at our table. "Looks like another installation from Morisette." Wordlessly Adin shifted over and handed him the first page of the transcript. Nico accepted the invitation, sat, and eagerly started to read, but Freud would have none of it. He strained at his leash and whimpered his protest. He wanted to go.

After a few minutes, Nico gave up on his attempt at appeasement. "Freud knows we're on our way to a hike, so he's raring to get going. Why don't the three of you come along? We usually go to Redwood Park. One of you can drive and I'll read the transcript on the way."

So that's what we did. Claire drove and Nico sat in the front with her. Freud, beside me, stuck his head out the window and sniffed the smells as they wafted by. "He's not too worried about developing an axiomatic theory for why the wind feels good on his face," said Claire smiling into the rearview mirror.

Nico read the transcript with pleasure. He nodded, and twice laughed in recognition. "What a great story," he said, finishing up just as we pulled into the entrance of Redwood Park.

But he didn't get into discussing non-Euclidean geometry right away. Instead he was content to talk about the park's history, its flora and fauna, and the geological forces that shaped the ridge we were headed toward. But most of all, it was the redwoods that interested him. "These are wondrous trees," he said in a sweeping gesture from the bottom to the top of a particularly gigantic specimen. "They grow to over a thousand feet and can live for two thousand years." He showed us the bark (resistant to lightning), the leaves (engineered for moisture retention), the shallow but strong root system (enabling the tree to get the surface moisture created by the coastal fog). "Also notice there are no nuts on these trees and that is why there are no animals or birds in a redwood grove." He was right. The grove was silent except for the rustling that the wind made near the treetops. "That is why the Native Americans considered these trees to be sacred. They said that the trees had the power to create peace which, in turn, leads to sublime thoughts," he said with his head uplifted, looking at the treetops. Even with my inherent skepticism toward anything even slightly "new age," I thought that the Native Americans were onto something.

"Vijay Sahni would have loved the redwoods."

It was Adin, not I, who said this. I knew he was right the instant he said it. Bauji would have liked walking this trail and looking at these trees. I asked Adin how he knew this. "I know his outlook. I think I understand what drove him."

I thought that the prime force in his life was mathematics; the Bauji I knew was driven by solving problems. I told this to Adin, but he shook his head disagreeing. "I think Vijay Sahni was driven by a quest for meaning. I think he wanted to understand what things are about and wanted to get there by a process he could feel certain about." Which of course were Adin's own two objectives. Was he transferring his personality to Bauji or was he on to something?

"What did you think of the transcript?" I asked Nico.

"It's fascinating. The judge and your grandfather were two remarkable men; each was committed to following the truth wherever it took him." He paused to remove Freud's leash so he could run free, and then he got serious about our discussion. To understand this last conversation fully I think we should step back for a moment.

"Remember, all this started with your grandfather presenting the premise that humans tend to believe in things blindly and that logical deduction was the only way we could possibly approach certain truth. But deduction turned out to depend on axioms, or postulates as Euclid called them. This was not a huge problem though, for having postulates did not compromise the notion of certainty as long as the postulates were self-evident. Vijay presented Euclid's geometry as a model of certain knowledge precisely because its postulates seemed simple and self-evident. But then this question of the fifth postulate started to make things a little murky. At first it was just the problem of its complexity. Euclid and many others accepted that the statement was self-evident, even though it was not simple. So they tried to prove it. But in their attempts to prove it they found that geometries that assumed the first four postulates and the falsehood of the fifth postulate were as consistent as Euclid's. Now this was a crisis because the self-evidence of Euclid's axioms should have allowed no room for any other consistent geometries. And yet, as we saw in this last transcript, Vijay constructed a proof for showing the consistency of geometries in which the fifth postulate was not true: the one in which he drew lines within a circle, but he didn't have to construct an artificial model. Non-Euclidean geometries are all around us." He threw out his long arms in a sweeping gesture to include the canyon in front of us.

"Really?" asked Claire. "How so?"

"The surface of the earth—or the surface of any sphere for that matter—is a non-Euclidean geometry," said Nico.

"What does a straight line even mean on the surface of the earth? I mean, you can't quite dig under the surface of the earth to get a line that is really straight," said Adin.

Nico made a fist to simulate a sphere. "Think of a straight line on a sphere as a great circle—a circle formed on the surface of the sphere by a plane passing through the center of the sphere. It's the circle that would divide the earth into equal parts: horizontally, vertically, or at any angle at all."

Claire nodded. "That makes sense. I guess if you take any two points on earth, say San Francisco and Tokyo, the shortest path along them is a great circle."

"Right," Nico said. "Any other path would be longer. The equator and all longitude circles are examples of a great circle. Now given this definition of a straight line on the surface of a sphere, which of Euclid's postulates are satisfied?"

The three of us had already started on this question. Only Freud was enjoying the sights around him.

A few seconds later Adin was shaking his head, "The first postulate fails. Euclid asked for a straight line, as in a 'unique straight line between any two points'. But if you take two diametrically opposite points, such as the

North Pole and the South Pole, it is possible to draw any number of great circles through those."

Nico nodded and laughed. "You're right, Adin. There are many great circles passing through diametrically opposite points. Notice that this does not happen for any other points; any points that are not diametrically opposite have exactly one great circle that connects them."

Adin brought his eyebrows together for a moment and then relaxed them. "That's right," he allowed.

"What if I interpret a point to mean point-pair—the point and its diametrically opposite pair?" I asked.

Nico smiled at the trick; I had anticipated his answer. "That'll make the first postulate work." He went on to show that the first four postulates worked just fine on the surface of a sphere. "The second postulate about indefinitely producing a straight line will work if we allow the great circle to retrace itself. The third postulate about drawing a circle from a given point and a given radius is true, although the definition of circle needs to be tweaked in this context, and the fourth postulate about all right angles is true exactly the way it is in Euclidean space. All of which brings us to the fifth postulate."

Adin was waiting for us to get there. "Nico, the fifth postulate does not make sense here. There are no parallel great circles. In fact, two great circles will intersect at exactly two points, so they can't be parallel."

"But that does give us a non-Euclidean geometry," said Claire. "Instead of assuming, like Lobachevsky and Bolyai did, that *two* parallel lines can be drawn through a given point, this model has *zero* parallel lines through a given point. So, in its way, this geometry is denying the truth of the fifth postulate."

"That is correct, Claire. Bernhard Riemann, Gauss' student of one of the greatest minds in mathematics, came up with this set of postulates, and you're right: this is indeed a third type of geometry different from Euclid's and Lobachevsky's."

"So there are many types of non-Euclidean geometries?" asked Adin.

"Yes there are," said Nico.

Adin walked looking at his feet, missing the wildflower-speckled plain that opened up on our right. Shortly we resumed our climbing, passing several thickets of madrone trees. In time Adin asked if Riemann's geometry was consistent just as Lobachevsky's and Bolyai's was.

"Indeed it is," said Nico. "In fact, you can use the theorems of solid Euclidean geometry to derive the axioms about great circles on spheres. So if this geometry leads to a contradiction, so must Euclid's."

It was an argument similar to what Bauji had shown the Judge.

Claire, as usual, was streaking ahead. "So the sphere's surface is a model for Riemann's non-Euclidean geometry. Is there a surface that is a model for Lobachevsky's version where there is more than one parallel?" she asked.

"There is," Nico said. "Imagine a concave surface such as a saddle or two funnels joined at their mouths. It's possible to show that these surfaces admit the first four postulates and have two parallel lines through a given point. Using this model you can conclude once more that Lobachevsky's geometry is as consistent as Euclid's."

Once again we walked in a silence that continued as we steadily climbed toward the summit. Along the way I looked at Adin, expecting him to be at least a little distraught at the news that there was not one, but at least two, non-Euclidean geometries that were every bit as consistent as Euclid's geometry. But he betrayed only a contemplative equanimity.

It was not until we reached the summit of the ridge and had spent a few minutes looking at the Bay Area spread out beneath us that he spoke again. "You know, Nico, the consistency of these non-Euclidean geometries is interesting, but it really does not create any kind of philosophical crisis. I admit that it's a little unexpected that these geometries even exist, but once you get used to the idea and let it simmer, you see that it isn't surprising that a curved surface has a non-Euclidean geometry." He stopped talking and I could almost feel his thoughts coming to order. "Think of a stretched rubber sheet. Draw a triangle on the sheet, towards one edge. Since the sheet is flat, the geometry is Euclidean and the triangle has exactly 180°." I nodded. We were good so far. "Now," said Adin, "put a heavy metal ball right in the middle of the rubber sheet. This will bend the sheet downwards and the triangle will become distorted—it will no longer have angles that sum to 180°." He picked up a curved stone from the dry streambed that was just adjacent to our hiking path. "In fact, Nico, if you take any curved surface, its geometry will not be Euclidean." He drew an imaginary triangle on the surface of the stone he had picked out.

"Adin, it depends on what the axioms are on that surface, but in general you're right," said Nico.

Adin was shaking his head. "Then the existence of these Non-Euclidean geometries is not surprising. When Euclid wrote up his axioms, he was thinking about space, not curved surfaces! He would not have been the least bit surprised if you told him that triangles did not sum to two right angles on the surface of a sphere, for he wasn't talking about the surface of a sphere, he was talking about space."

Claire pointed out that Lobachevsky thought that his geometry, not Euclid's, was the geometry of space.

"That couldn't be," said Adin. "It does not make sense."

So here we were, finally, at the real issue. Bauji had thought that Euclid's methods provided certainty about the nature of space around us. He had built a geometry for this space. Lobachevsky, Bolyai, and Riemann had shown that Euclid's geometry was not the only possible geometry, and that their geometries were as consistent as his. But what was the actual geometry of space?

"There are two things going on here," said Nico. "One is the creation of a logical deductive structure starting from some definitions and axioms. All geometries we have seen—Euclidean and non-Euclidean—do this successfully. The second is the question you are asking, Which geometry is the actual geometry of space?"

Adin reacted to the second question. "You're telling me there is a chance that the geometry of space is non-Euclidean?" he asked, his eyebrows traversing new heights.

"Absolutely there is—but it is a question that mathematics is silent on."

"But that is *the* question," countered Adin. "Why is mathematics silent about it?"

Nico gave him the party line. "Mathematicians care about axioms and their logical implications. The nature of space and which axioms it follows is a question for physics."

"That," said Adin, "is an absurd position."

Claire, for the first time in many weeks, agreed with Adin. "We can't ignore that question, Nico. Do you really believe that the fifth postulate can fail in space?"

"It could," said Nico. "I don't have intuition for how it may fail, but I admit the possibility that it could fail. We know that the surface of the earth is a sphere, which is why we have no trouble visualizing a non-Euclidean geometry on it. Maybe it's the same with space. Maybe it curves in a way we don't understand. I've read some things about this but I don't understand it fully, so I'm not going to talk about it."

That's the way Nico was. His understated approach, even playfulness, made it easy to forget how precise he was about everything he taught. I later found out that Nico did know quite a bit about theories of the nature of space, but he didn't have all the details, so he refused to say anything at all.

"What I can tell you," he said as we started our descent back, "is that there are wonderful parallels between the fifth postulate and the Continuum

Hypothesis." Nico had made this observation before, and ever since he had made it, I was curious to know more.

"It's like this," he said, keeping one eye on Freud, who was making friends with a gorgeous chocolate Lab up the trail. "Cantor developed his theory without starting with axioms. In time, some problems began to emerge and the only way to address those problems was to develop the theory based on intuitive axioms. This task was successfully taken up by a mathematician named Ernst Zermelo. In some ways Zermelo was the 20th century version of Euclid. Remember, Euclid did not prove his propositions; they were known before him but had never been rigorously proved. Geometers such as Pythagoras had demonstrated the propositions by relying on intuitive notions of lines and points. They had proceeded without worrying too much about definitions and axioms. But then, quite possibly in response to paradoxes such as Zeno's, Euclid came on the scene and provided a solid foundation for the geometry. A similar unfolding occurred in set theory, where Zermelo provided the foundations after Cantor had developed his theory of infinity based on intuitive notions of sets. When paradoxes started to emerge, Zermelo stepped in and provided the definitions and axioms that prevented the paradoxes from arising."

Claire was having difficulty accepting that Cantor's elegant theory could have paradoxes hidden within it.

"They're really easy to explain," answered Nico. "You remember that the cardinality of the power set is greater than the cardinality of the underlying set. So if S is a set, Cantor had shown that the power set [S] > S. Now let's define U to be the set of all things in the universe. U contains all things, ideas, sets, subsets. Every last thing that exists is a member of U." Nico was flowing. Freud was long forgotten and several large redwoods had passed by unnoticed. "Here's the paradox. What is the cardinality of power set [U]?"

Claire was with him. "I see the problem. The power set [U] is supposed to have a greater cardinality than U, but U is the set that contains everything. So a set couldn't possibly be greater than U."

"Exactly!" said Nico. "There's the contradiction. U can't be the set of everything and yet be smaller than another set."

Adin wanted resolution. "So Zermelo solved this problem by formalizing the subject with axioms and definitions?"

"Yes, exactly. He introduced the definitions and axioms of set theory and under his axioms all of Cantor's theorems stayed intact, but the paradoxes were banished and order was restored."

"All this is fine, but what is the connection between the fifth postulate and the Continuum Hypothesis?" I asked.

"I'm getting to that, Ravi," said Nico. "It turned out that it was possible to prove that the Continuum Hypothesis is independent of Zermelo's axioms. It can be neither proved from the axioms nor disproved. In this sense, it is like the fifth postulate. The fact that Euclidean geometry exists, and is apparently consistent, means that you can't disprove the fifth postulate from the first four; and the fact that non-Euclidean geometries exist means that you cannot prove the fifth postulate from the first four. In other words, the fifth postulate is independent of the first four."

This immediately raised an intriguing possibility. "Does that mean that just like non-Euclidean geometries, there are versions of set theory in which the Continuum Hypothesis is not true?" I asked.

"Indeed there are," said Nico. "The Continuum Hypothesis and the fifth postulate are independent of the axioms before them. You could assume them to be true and you get one theory, or you could assume them to be false and you get another theory."

Adin was the most agitated I have ever seen him. He shook his head vigorously in disagreement. "Wait a minute," he said in a shrill voice. "The fifth postulate is true or it is not true. Either a triangle in space has angles that sum to 180° or it doesn't. Similarly, the Continuum Hypothesis is true or it is not true. You either have a set that is bigger than the natural numbers but smaller than the real numbers, or you don't. It doesn't make any sense to try to build these theories on axioms that are untrue, even if by some miracle these theories appear to be consistent. I mean, truth exists, doesn't it?"

Nico nodded. For a minute he looked tired and closer to the age he actually was. "This is the issue in all of mathematics, quite possibly it is one of *the* questions in all of human thought." Then he sighed and made a decision. "We need to talk a lot more about this stuff before you draw any general conclusions about life or about the questions Vijay and the Judge are thinking about. It would be a shame to sketchily understand the underlying mathematics and base your entire world view on incomplete information. We need to sit down and carefully talk through the underlying issues. Come over to my place and we'll do it—tonight."

CHAPTER

1 2 3 4 5 6 7 8

IN THE CAR, NICO ASKED ADIN if, like Bauji's snake girl, something specific had triggered his interest in "knowing about knowing." Adin had paused, considering his answer. "I think it happened when I read 'The Red-Headed League'," was what he came up with.

None of us needed to be told that "The Red-Headed League" was a Sherlock Holmes adventure—one of the quirkier exploits of Conan Doyle's fictional detective. It turned out that Claire, Nico, and I had at various points in our lives been avid Sherlock Holmes fans, but none of us could recount the details of this particular episode the way Adin could. He told the story in a style that Doyle himself would have approved of – building slowly, with the right amount of color and context, the crucial clues provided up front, sitting there in plain sight for all of us to process along with the masterful Holmes. In places, Adin even acted out the dialogue among the characters, his voice changing with pitch and tone according to what he thought the speaker should sound like. His natural voice he saved for Sherlock.

In "The Red-Headed League" Adin described Holmes being consulted by Mr. Jabez Wilson, a widower and a somewhat unprosperous pawnbroker by trade. Wilson shows Holmes and Watson an advertisement in one of London's papers offering a lucrative employment opportunity for a qualified member of the Red-Headed League. The only membership criterion for league membership is a "real bright, blazing, fiery" shade of red hair, which Wilson clearly possesses. He has been shown the advertisement by his assistant, a Vincent Spaulding, who has been working for Mr. Wilson on half-wages so he can learn the pawn-brokering business. Spaulding is smart and hardworking and his only drawback, according to Mr. Wilson, is a preoccupation with photography, which keeps him in the darkness of the cellar at all hours. But this is a small flaw in the scheme of things and it is clear that before the episode of the Red-Headed League, Wilson, Spaulding, and a young part-time cook (age 14) had been living their days in relative peace. Since the pawnshop is located in Wilson's home, he was never required to travel anywhere.

Some eight weeks previous, Spaulding had shown Wilson the offer of employment from the Red-Headed League, insisting that Wilson apply for the vacancy.

"From all I hear it is splendid pay and very little to do," Adin said in a low pitch he imagined appropriate for Spaulding. Wilson found out that Spaulding was right on both counts. After being chosen from among dozens of red-headed men from all parts of London (none had hair quite as bright red as his), Wilson was asked merely to transcribe the Encyclopedia Britannica. His employers put no conditions on his work other than that he was not allowed to leave the League office during work hours for any reason whatsoever, or he risked losing his employment immediately.

And so Mr. Wilson didn't leave. He showed up every morning, worked diligently in the agreed-to hours, and was paid his salary on Saturday. All this suited him just fine because the extra money was a big help to him. But when he reported to work at the beginning of the ninth week, he found the office locked, and a sign on the door stating simply that the Red-Headed League had been dissolved. There was no forwarding information for his employer. This is when Mr. Wilson decided to come and consult Sherlock Holmes.

Holmes is charmed by the puzzle—a "three pipe problem," he calls it. After meditating awhile he asks Watson to come with him to Mr. Wilson's home (and pawnshop). On inspection, Holmes lists the neighboring establishments: "the tobacconist, the little newspaper shop, the Coburg branch of the City and Suburban Bank, the Vegetarian Restaurant, and McFarlane's carriage-building depot." At the pawnshop, Holmes examines the sidewalk in front of the building very closely, beats his cane upon it vigorously, then makes a point of speaking to Wilson's assistant, Vincent Spaulding. He cryptically asks whether Watson observed the knees of Spaulding's trousers. They are quite wrinkled.

"You now have all the clues that Holmes had to solve this puzzle," Adin told us dramatically. Claire, Nico, and I had listened to Adin's story in detail, each of us nodding at various times in recognition. But none of us could quite recall the nature of Holmes' solution. And our problem-solving skills were on the hunt!

"Here's how Holmes went about it," said Adin, enjoying the moment. "It was immediately obvious to him from the first that the only possible object of this rather fantastic business of the advertisement of the League and the copying of the Encyclopedia 'must be to get this not over-bright pawnbroker out of the way for a number of hours every day'. 'It was a curious way of managing it, but really it would be difficult to suggest a better,' is what he told Watson over a whiskey and soda after solving the crime. The entire ruse was suggested by the striking color of Wilson's hair. One accomplice became the assistant and incited Wilson to apply, and the other rented offices and pretended to be the employer!"

But Claire wasn't sure yet. "He wanted to get Wilson out of the house, but why?"

Adin laughed. "See, Claire, from the time that Holmes heard of the assistant having come for half-wages, it was obvious to him that he had some strong motive for securing the situation. There was nothing of value in Wilson's shop (Wilson was poor), nor were there any women involved (he was a widower), so Holmes deduced that the reason must be outside the pawnshop."

"The bank!" said Nico, using the same tone he used to describe Cantor's Aha moments. "Didn't you say there was a bank nearby?"

"Exactly," said Adin. "Holmes deduced that the assistant must have been doing something in the cellar—something which took many hours a day for months on end. What could it be? Holmes could think of nothing save that he was running a tunnel to the bank nearby. The assistant's wrinkled pants and stained knees confirmed his suspicions. Beating his cane on the street showed him that the cellar extended in the back towards the bank. Knowing all of this, he was able to set a trap for the criminals and catch them red-handed."

"And not red-haired," I remarked. Claire laughed.

It was a nice solution and for a time we all went silent, looking at the chains of logic that Holmes had used to solve the mystery. "Pretty cool," said Claire. "It fits together very well."

"That's what struck me about it," said Adin nodding his head. "There is an inexorable logic, the conclusions are supported by underlying reasons, and nothing is arbitrary. Even as a boy, I realized that Holmes' reasoning was qualitatively different from that of those around me—he drove to his conclusions without passion or prejudice. He didn't argue a point of view merely because he was predisposed that way. He looked at the facts and made logical deductions from them. It seemed to me then, as it still does to this day, that Holmes' way leads to a reliable truth, whereas the people I saw around me proceeded in their lives with a mixture of guesswork and unanalyzed emotion. So Holmes was my snake girl moment. He showed me the path to certainty."

The word "certainty" got to me. If Euclid's system could turn out to be susceptible to doubt, surely Holmes' was not immune either. Nico had been thinking along similar lines. "Adin, I agree Holmes' reasoning seems airtight, but let's examine this in terms of the mathematics we've been learning. Holmes' first conclusion was that the league was a sham. How does he conclude this? Is it based on any axioms?"

"Holmes does not explicitly refer to any axioms," said Adin.

"You're right," said Nico, "he does not. Strictly speaking, Holmes is not following the axiomatic method at all. In a true axiomatic system we would have a set of axioms and the theorems would follow inexorably through a series of logical steps. Vijay Sahni spoke of just such a system as being deductive in the last transcript. An American philosopher, who wasn't a bad mathematician himself, Charles Sanders Peirce, observed that the inductive way of knowledge ascends from fact to law whereas deduction, applying the pure logic of mathematics, reverses the process and descends from law to fact. The method Holmes is following is not really deductive—he has a set of facts from which he infers an explanation."

This did not seem to satisfy Adin, who seemed to have rethought his example in light of what we had been speaking about. "But there is an implicit axiom—or perhaps it's a set of axioms of what Holmes considers to be the expected band of human motivations. Expected motives may include a quest for power, wealth, love, or knowledge. Only Holmes could provide the specific details, but let's say for a moment he had such a list and this list constituted his implicit axioms. If he finds something that lies outside that band of expected motivations, then Holmes would conclude that there must be some other explanation for the action. A league that chooses an employee solely on the basis of his hair color and asks him to perform a useless task is outside his expected band of human behavior and so there must be another explanation. So Holmes examines the result of the action and concludes that the only possible motive was to achieve what was in fact achieved: Wilson was predictably outside the shop for several hours every day. In a sense you could argue that this is a theorem derived from Holmes' axioms of human behavior. Holmes then asked why the criminals could possibly have wanted Wilson to be outside the house. None of the possible explanations within the house fit with Holmes' behavioral axioms—there is no money to steal nor are there women to have romantic interludes with—, so the explanation must lie outside the house—a theorem that follows deductively from the axioms."

Claire objected, shaking her head, "I can't see how Holmes could explicitly state all his axioms about human behavior. And even if he could it would be very difficult to see how they would be as indisputable as the Euclidean axioms."

"Right on both counts, Claire," said Nico, "The distinction I made earlier is important. When we infer explanations from facts we arrive at a plausible explanation. This, in fact, is how much of science works. But the inferences are usually approximations to the truth, liable to be supplanted by better explanations. On the other hand, if you consider the axioms of Euclidean geometry, they are a complete deductive system. There could be alternate systems but they do not supplant Euclidean geometry. However, the process Adin describes may well be how the first geometers approached geometry. It was Euclid's greatness that he could formalize the axioms in a workable fashion. This attempt at formalization would have followed a process of inference. Maybe many different sets of axioms may have been guessed at, and some wrong axioms may have been tested, but thanks to Euclid's genius a true deductive system was developed in the end. But a deductive system is only as good as its axioms, and it is completely valid to ask if the axioms are true of the space we live in."

"I absolutely refuse to believe that there is uncertainty in Euclid's axioms," said Adin, echoing my grandfather's sentiments. "If there is a flaw in them, then we can never be sure of anything."

Nico said nothing, and from the backseat I couldn't make out the expression on his face.

• • •

Everyone was hungry by the time we got back to Nico's place and so we gladly accepted his invitation to stay for dinner. "If you guys help with slicing and dicing, we'll be eating in 30 minutes," he promised. So we worked with gusto. Claire took charge of the salad, Adin handled the basil leaves for the yoghurt sauce, and I chopped vast amounts of garlic into little cubes.

"Ravi, if you chop the garlic too fine it will make the dish bitter. Aim for slightly larger pieces," said Nico.

Nico cooked liked he did everything—with passion and enthusiasm—and it showed in the results. We feasted on the Greek salad (with olives), lamb marinated in a garlic sauce, and homemade pita bread. The food was good enough to induce concentration and the four of us ate in silence. Dessert consisted of Baklava and tea, and the Baklava too was homemade. "The stuff they sell in stores is too dry," Nico explained.

After dinner we went to the study and it immediately reminded me of what Bauji's room used to look like. There were mountains of books and piles of jazz CDs, but Nico immediately drew our attention to a slim, well-used volume that lay open on the middle of his desk. It was titled *The Continuum Hypothesis* by Paul Cohen. "I first read this book about 20 years ago," said Nico, "and I still think it's a classic. After I read it, I understood the power and the limitations of the axiomatic method." I recalled Nico saying that Paul Cohen was the man who finally showed that the Continuum Hypothesis is independent of Zermelo's axioms, but I didn't yet fully understand the context or implications of the statement.

Nico sat on what was clearly his favorite chair by a window that looked out into his backyard. Claire and I sat on the carpet and Adin took the little chair by the desk. We settled in, getting comfortable. This was going to be fun.

"Before talking about anything new it always helps to look back on where we've been," Nico began. "I know you have all read about the Pythagoreans. Remember, these guys were around some 300 years before Euclid and they did some pretty great geometry, although none of it was axiomatized. They relied on intuitive notions of a point and a line. Then Euclid came along, and perhaps in response to some paradoxes, came up with the five axioms which you read about in the Sahni–Taylor transcripts. These axioms were constructed to match our intuitive geometric notions but did not rely on intuition to prove any theorems.

"In set theory Cantor plays the role of the Pythagoreans. He founded the discipline of set theory and found some wonderful theorems by relying on the intuitive notions of a set and belonging. The role of Euclid was played by Zermelo. Recall that in response to some paradoxes in set theory, Zermelo axiomatized the discipline. Once again, he removed the need for intuition. If you believed the axioms, you had to believe the theorems.

"Now look back at what happened after Euclid. For aesthetic reasons, people tried to prove that the fifth postulate followed from the first four. As we've seen, these efforts failed repeatedly, and finally led to the creation of non-Euclidean geometries. The question then arose whether these geometries were consistent, or at least as consistent as Euclid's geometry. Consistency just means that there are no contradictions contained in the subject.

Now, remember that by and large people agreed on the first four postulates. If you added the fifth postulate you got Euclidean geometry and if you added the negation of the fifth postulate you got non-Euclidean geometries. As we have seen, it was shown by considering simple models that the non-Euclidean geometries are as consistent as the Euclidean geometries.

"But what about the consistency of Euclidean geometry itself? No one even thought to try to prove that there weren't some hidden paradoxes in Euclidean geometry. Frankly it's quite understandable that no one asked this question. Euclid's geometry was the geometry of our experience; to suspect a contradiction in it was tantamount to suspecting a contradiction in human experience. It was not until other geometries were established that the question even arose. When it did arise, the question was answered by David Hilbert. He was able to show that Euclidean geometry was as consistent as elementary algebra. Hilbert showed this equivalence by associating each point on the Euclidean plane with a pair of numbers: the familiar x and y coordinates. Then, with each circle, line, and square on the Euclidean plane, there is a unique algebraic relationship or equation that corresponds to that particular circle, line, or square. Thus if there is a contradiction in geometry one could carry through that contradiction to elementary algebra. So non-Euclidean geometry is as consistent as Euclidean geometry, which is as secure as elementary algebra.

"Here's the interesting thing: the role of the parallel postulate in geometry is roughly played by the Continuum Hypothesis in set theory. As we've seen before, Cantor's set theory, axiomatized by Zermelo, leads to the formulation of the Continuum Hypothesis, which states that there is no cardinality between the set of natural numbers and real numbers. As I've told you before, Cantor and many others tried to prove the Continuum Hypothesis, but they all failed. This was analogous to the attempts of geometers to prove the fifth postulate from the first four. But Kurt Gödel, an Austrian mathematician, was able to show that if we take the Continuum Hypothesis as an axiom and add it to Zermelo's axiom set, then the new expanded axiom set is as consistent as the original Zermelo axiom set. In other words, if there is a contradiction in the expanded axiom set (Zermelo + Continuum Hypothesis), then the contradiction must already be present in the Zermelo axiom set. Notice that Gödel did not prove the Continuum Hypothesis; instead, he was able to show that the Continuum Hypothesis could not be disproved using Zermelo's axiom set. If there was such a proof (disproving the Continuum Hypothesis), then adding the Continuum Hypothesis as an axiom would immediately lead to a contradiction since one would have the Continuum Hypothesis and its negation as true statements, which is a logical impossibility."

We had listened to Nico without interruption. Outside, a fading dusk had given way to darkness, and after a while we couldn't see his face but could only hear his voice. I thought through what Nico had just said and confirmed in my head that he was right. If someone disproved the Continuum Hypothesis he would have proved the negation of the Continuum Hypothesis (i.e, that there is a cardinality between the naturals and the reals). Adding in the Continuum Hypothesis as an axiom would be an immediate contradiction.

"So what Gödel did was analogous to what Hilbert had done," Nico continued. "He proved the relative consistency of expanded set theory—Zermelo + Continuum Hypothesis (CH)—by showing its equivalence to a simpler system, in this case the basic set theory of Zermelo.

"What's fascinating is that this analogy between set theory and geometry continues, and even gets stronger. Remember that non-Euclidean geometry arose by assuming the negation of the fifth postulate. Similarly, assuming the negation of the Continuum Hypothesis leads to another kind of set theory."

"Is there a proof of the relative consistency of this new set theory?" asked Adin.

"Ha! You guessed it Adin," said Nico. "It was Paul Cohen," he said, tapping the cover of the book in front of him, "who was able to prove that the negation of the Continuum Hypothesis when added to Zermelo's axioms leads to a set theory that is as consistent as basic set theory. This was analogous to proving that non-Euclidean geometries created by assuming the negation of the fifth postulate are as consistent as basic geometry (geometry from the first four Euclidean postulates) and Euclidean geometry."

"Wait a minute, said Adin, shaking his head. "Hasn't Cohen proved that it is impossible to prove the Continuum Hypothesis, just like Gödel had shown that it is not disprovable?"

Nico nodded his head. "Exactly! These two statements amount to saying that the Continuum Hypothesis is neither provable nor disprovable from the basic Zermelo axioms."

"Just like the fifth postulate is neither provable nor disprovable from the other four postulates," I observed.

"I see," said Claire. "If the fifth postulate were provable, then the non-Euclidean geometries would be inconsistent, and if it were disprovable—that is, its negation was provable—then Euclidean geometry would be inconsistent."

But Adin was still not buying it. "The objection I made at the end of our hike still seems valid to me," he said. "Either space is Euclidean or it's not. Either there is an intermediate set between naturals and reals or there isn't. So what if these alternative geometries and set theories are consistent? That does not make them real."

Right then I realized something I thought to be interesting, even important. "Adin, I suspect that from a purely mathematical viewpoint non-Euclidean geometry is as valid as Euclidean geometry. What the actual geometry of space happens to be is a separate question. Mathematics is concerned with axiomatic systems and their relative consistency. It is similar to set theory. In one flavor of set theory you get to assume the Continuum Hypothesis, and in another flavor you get to assume its negation. Both set theories turn out to be consistent. Here I don't even know if it makes sense to ask which set theory is real. Is there a set between the naturals and the reals? I don't know. If it did exist, it would only exist in our minds, wouldn't it? All this stuff is in our minds. What exists out there is a different question; perhaps the answer is unknowable."

We were getting into some interesting territory and I wanted to be sure that I was clear on how we got there. Mentally I constructed a picture:

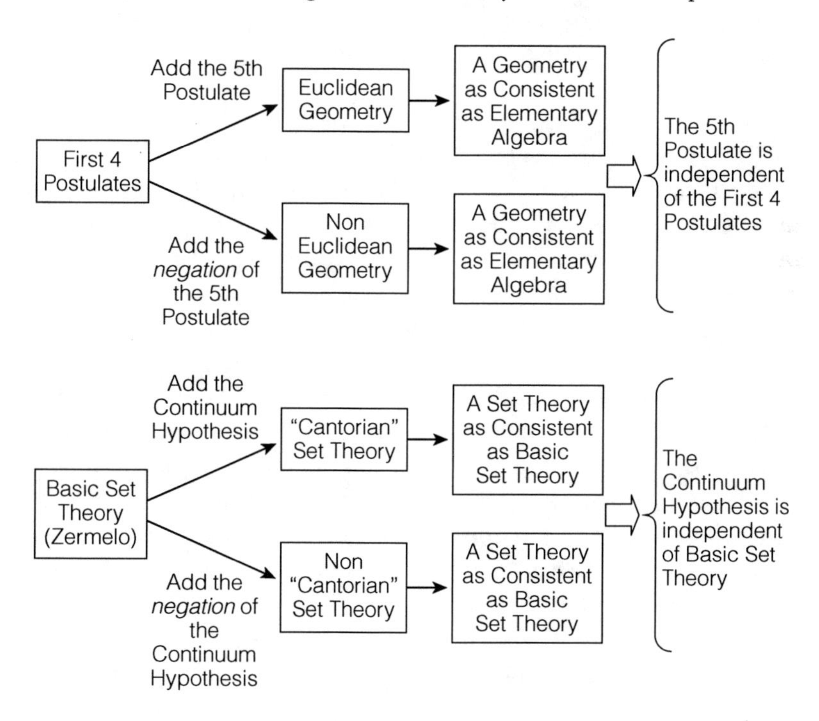

But Adin wasn't satisfied. "What is definitely not just in our minds is the nature of space. Either, as our senses scream, a point can have exactly one parallel line through it, or it can have more than one. One possibility must be correct, and the other cannot be. And saying that these axioms are independent of one another, almost as if it's an arbitrary choice, is completely repulsive to me. I happen to be quite sure that if we could determine the true nature of space we would find it to be Euclidean. Also, maintaining that mathematics does not address the nature of space is a completely unsatisfying position. Someone should address that question."

"Someone did," said Nico. "You guys need to read up on Gauss, Riemann, and Einstein. Those guys did address that question."

"And what did they say?" asked Adin.

"You will have to find out for yourself," said Nico. "It's not my area of expertise," he said, walking over to the lamp and switching on the light.

• • •

JOHANN CARL FRIEDRICH GAUSS MAY 1854, GÖTTINGEN, GERMANY

Farkas Bolyai's letter got me thinking today. He asked, "What is the single problem that has most troubled the greatest mind that ever lived?"

With his usual dramatic flair he calls me "the greatest mind that ever lived." I have become accustomed to ignoring such flattery from him. But his question is a good one for it provokes introspection. Indeed, what has been the most vexing and the most important problem of my career?

The answer must be the problem of Euclid's fifth postulate.

Not because the mathematics of parallels is more difficult than other mathematics I have done, but because the problem forces us to go outside mathematics in a manner that no other problem has ever done. It forces us to re-examine some of our most cherished assumptions, and ultimately it makes us think about what we know and how we know it.

After 30 years of intensive work, I am now convinced that the assumption that the fifth postulate is false leads to a peculiar geometry, which is quite distinct from the Euclidean, and quite consistent. For myself I have developed it quite satisfactorily. All my attempts to find a contradiction, an inconsistency in this non-Euclidean geometry have been fruitless, and I am convinced that there will never be such a contradiction. I have never published anything on this subject fearing the clamor of the mediocrities who will not be able to consider anything other than the geometry they live with.

Many years after I discovered this curious geometry, I received communications from Farkas' son Janos and a Nikolai Lobachevsky from Russia, rediscovering many of the theorems I knew. I commend these men for their insight and courage (and for saving me the trouble of having to publish my own results in this area), but I fear that despite my encouragement their work will remain largely anonymous. That is the nature of the world we live in.

Given my belief that non-Euclidean geometry is free of contradictions, I began to grapple with the next natural question, What is the true geometry of space? The simplest way to decide this critical question is to measure the sum of angles of a triangle. If the sum is exactly 180°, then the geometry of space is Euclidean. If, however, the sum of the angles is less than 180°, then the geometry of space must be non-Euclidean. The only complication is that to notice any difference in the sum of the angles we must consider large triangles in space. For smaller triangles, we have no instruments that could detect the very small differences in the sum of the angles.

I decided to resolve the question with an experiment that actually measures the angles of a large triangle. It would have been no help trying to draw a large triangle on the surface of the earth. It is well known that the earth is a curved surface and the geometry of any curved surface is non-Euclidean (Euclid himself would have known this). No, the triangle had to be drawn in space. So one dark night I arranged for fires to be lit over the three peaks of Hohenhagen-Inselberg-Brocken and measured the sum of the angles in this large triangle (at the time I had other reasons for this experiment as well).

Unfortunately the results were not conclusive. The triangle did not display any angular defect but I do not know if this was because space is really Euclidean, or if my triangle was far too small to measure any defect that might exist. The problem remained unsolved.

Now, as my health is failing, I'm turning to the best mathematician of the next generation to shed light on this problem. His name is Bernhard Riemann. He is a student of mine and his work on functions of a complex variable is the work of a genius—and I am not one to use that term loosely. Riemann is scheduled to deliver his presentation paper in a few weeks. All doctoral students are required to submit a prioritized list of three topics for their presentation papers. Most students are given their first choice, and in a few rare cases they are awarded their second choice. But for Riemann I picked his third choice. He may not be too pleased at that, but the importance of his third topic made my decision automatic. He is scheduled to present a paper titled "On the Hypotheses that Lie at the Foundations of Geometry."

• • •

BERNHARD RIEMANN JUNE 9, 1854, GÖTTINGEN, GERMANY

In his wisdom, Herr Professor Gauss has asked me to present on the topic I know the least about. I had put down my main research interests as my first two choices for my Habilitationschrift, confident that I would be well prepared for either. Yet my assigned task was to be to investigate the foundations of geometry—something I have thought about sporadically; certainly not something I had mastery over.

So I spent the last several weeks locked up in my quarters developing an entire theory from scratch. Up until the morning of my presentation I was not sure that everything would come together, yet it did quite nicely. My presentation was well received. I believe even Gauss was impressed!

My fundamental idea was to strive for unity amongst the various geometries that exist today. Euclidean geometries work on flat surfaces; Lobachevsky, Bolyai, and Gauss have developed a geometry that works on hyperbolic surfaces; and I myself have a geometry that works on spherical surfaces. But underlying all these geometries is Gauss' notion of curvature. He had defined the curvature of flat space to be zero, the curvature of a sphere as positive, and the curvature of a hyperbola as negative. To unify these various curvature-driven geometries, I put forward the notion of a general metric to measure the distance between any two points.

After some thinking that I will not include in these notes, I was able to derive a generalized distance metric that can be made to work for any curved surface even if its angle of curvature changes from point to point. If ds is the distance metric, its value is given by the following equation:

$$ds^2 = g_{\mu\nu}\, dx_\mu\, dy_\nu,$$

where μ and ν are indices that take the values 1 and 2 and $g_{\mu\nu}$ is the metric tensor. This metric is useful in describing the curvature of any surface, even if it is not uniform and even if the curvature changes with time.

Of course these techniques do not shed light on the actual nature of space, I am unable to say if space is flat, as everyone believes, or non-Euclidean as Gauss and a few others are beginning to suspect.

Perhaps one day someone will use my techniques to make progress on that question.

• • •

ALBERT EINSTEIN PERSONAL NOTES, NOVEMBER 18, 1915

All those years my friend Grossman and I looked for mathematical tools that would allow us to push forward our ideas on gravitation and space. After a lot of trying, I finally found what we were looking for. It is a metric thought up by the mathematician Bernhard Riemann, and it has been with us for about 60 years!

Riemann's metric solves many of the difficulties I had been grappling with. I can simply take the equation

$$ds^2 = g_{\mu\nu}\, dx_\mu\, dy_\nu$$

and let μ *and* ν *take the values 1,2,3, and 4, and my theory starts to fall very nicely into place.*

It now seems clear to me that heavy objects "curve" the space around them. **Space need not have a uniformly Euclidean geometry**. *Gravitation works, not by applying force from a distance, but because objects "fall" along this curvature of space-time.*

If I'm correct, it should be possible to observe the bending of starlight caused by the sun. In fact, my calculations predict that a light ray from a star just grazing the sun should bend by 1.75 arc seconds. But it is impossible to take such measurements without a solar eclipse. I certainly hope that someone understands the importance of taking the measurements and avails himself of the next opportunity.

• • •

Things started to come to a head in the next transcript.

NOVEMBER 9, 1919

Vijay Sahni vs. People of New Jersey

OFFICIAL COURT DOCUMENT

JUDGE TAYLOR: Read this.

[Note from Court Reporter: Judge Taylor handed the prisoner a clipping from a New York newspaper. It is reproduced below. As the prisoner read through the clipping the judge paced the room. He could not keep still and he seemed to be under great emotional stress.]

REVOLUTION IN SPACE!

Morisette — Newtonian ideas eclipsed. Space is "curved"

The old order of physics that has stood unchallenged since the time of Newton has given way. Observations made in the far-off Principe Islands in West Africa have confirmed one of the most startling predictions of Albert Einstein's theory of general relativity: light does not always travel in straight lines but rather bends due to gravity in the proximity of massive objects such as the sun.

Astronomer Arthur Eddington, who led the expedition, announced the results of these observations at the Royal Society. The observations could only have been made during a total eclipse of the sun, which allows photography of stars close to the sun. The solar eclipse of May 29, 1919 was considered suitable for verifying the predictions and Eddington's expedition sailed for the Islands from England in March. According to Eddington, the photographs of the eclipse have demonstrated the effect of gravity on light rays passing close to the sun.

Eddington said that this effect was not predicted by Newton's theory of gravitation, thus clearly demonstrating the superiority of Einstein's theory. Einstein had proposed his theory of general relativity in 1915. Eddington explained that the general theory of relativity, apart from the three dimensions of space, considers time itself to be an extra dimension. In this view of the universe, massive objects such as the sun have the affect of bending this four-dimensional space-time in their proximity. As a result, while light, as in the Newtonian universe, continues to travel the shortest distance between two points, this shortest distance is no longer necessarily a straight line due to the curved nature of space-time near massive objects.

In order to verify Einstein's predictions, Eddington had to wait for a particular set of circumstances. Seen from the earth, the sun moves against the backdrop of stars. It is theoretically possible to photograph stars in the sky when the sun is far removed from them and then photograph the same cluster again when the sun passes close to them. In the second set of photographs the light from the star on the way to the earth passes very close to the sun and thus, according to Einstein, is subject to the bending effect of gravity. When the two sets of photographs are compared, the position of the stars, if Einstein is right, do not match.

But, Eddington further explained, under ordinary circumstances it is not possible to photograph stars very close to the sun because of the sun's brightness. It is only during a total solar eclipse, when the sun's brightness is sealed off from the earth by the moon, that it becomes possible to carry out the proposed observations.

[Note from Court Reporter: Once started, the conversation below had frequent pauses. Unlike previous instances, both the Judge and the prisoner spoke in soft tones and pieces of the conversation were not intelligible. Court reporter's requests for repetition were ignored.]

JT: I am not sure I understand.

VS: I don't know what to think anymore.

JT: What do you mean?

VS: Well. This seems to say that Euclid's fifth postulate may well be false.

JT: What?

VS: I have heard that nothing in Einstein's theory requires that the geometry of the universe be Euclidean. The Newspaper report seems to suggest Einstein is right. And now I ask myself, If deductions from Euclid's axioms lead to propositions that are not only uncertain, but also possibly untrue, then how can we know anything at all?

[Note from Court Reporter: Judge Taylor's statement was unintelligible.]

VS: Perhaps the people of Morisette were correct to prosecute me after all. I've based my life on a system that could be deeply flawed. At least, I have no certainty that I am right. Perhaps you religious people have access to a type of knowledge I just don't understand. Maybe I should support your decision recommending that my case go to trial after all.

JT: Then, Vijay, you have not understood the reasoning behind my recommendation. Since the time you showed me Euclid's axioms I have believed in this deductive approach of yours. My argument has never been with your methods, only with your choice of axioms. I could never understand your refusal to accept my axiom of the existence of the Almighty. I felt that it was self-evident and that it was possible to derive many lessons on how to live a good life from that one axiom alone. So, contrary to what you seem to believe, I have come to believe in your axiomatic methods.

VS: Judge, as you have seen, no axiom is safe. If no axiom is safe, then no deduction is possible. And what use are my methods then? My everyday certainties—certainties that I thought lay behind my passion for mathematics—do not seem to stand on very firm ground now.

JT: I must say that my convictions do not stand on firm ground either. If it is possible for Euclid's fifth postulate to be false, then perhaps any postulate could be false.

VS: Even your postulate about God?

JT: *[Unintelligible]*

• • •

That night was the first (and last) time I ever saw Adin noticeably and unquestionably drunk. We had gone down to the Coffee House after reading the transcript about Eddington's discovery, and instead of his usual raspberry lemonade, he'd ordered a shot of gin. I knew he was pensive and troubled, so I resisted the temptation to ask him what he was thinking. It was not until after his third drink that he looked toward us and seemed ready to talk.

"So even Euclid's axioms could be wrong," he said.

Claire, who had been shaking her head in disagreement as we read the last transcript, had a different take, "*So what* if Euclid's axioms don't apply to space? What I'm not getting is why this is such a big deal. The theorems are still true in that they follow from the axioms. I understand that it's surprising, but it's certainly not something to get depressed over." Her tone and pitch were markedly raised and she sounded like the mother of a delinquent teenager.

Adin shook his head. "You're missing the point Claire," he said looking at her. "This whole thing is about being certain about ideas."

"I understand," said Claire, sounding like she didn't understand at all. "But we do have certainty for all practical purposes, Adin. There are clear laws in life that work. Now, maybe you can't have absolute and abstract certainty that these laws are true, but that's the way it is; that's the way life is."

Adin looked directly into Claire's eyes. Claire, not usually one to back down, returned his gaze.

And then most unexpectedly Adin raised his voice. "Don't you understand that without certainty there can be no meaning?" he shouted. "There would be nothing left, nothing."

We were silent for several minutes. When Adin spoke again, it was in his normal tone of voice. "When I started studying mathematics I thought I could finally behold real and true certainty. But as I read the transcripts, I found that the only way to certainty was by picking certain axioms. Without axioms things were rootless. You needed axioms; Euclid showed that. I persuaded myself that it was okay to have axioms as long as you picked axioms you were intuitively certain of—and Euclid and even Zermelo seemed to have successfully done this. Then we learned that no matter what axioms you picked, there would be some questions, such as the Continuum Hypothesis, that would be impossible to answer. I saw this as a blow, but not a fatal blow. It showed that from a given set of axioms one could only achieve partial certainty. Not great, but still the dream of certainty was intact. Next we saw that the axiom we select is itself a matter of choice: one could use the Continuum Hypothesis as an axiom and get one type of set theory, or one could choose to use the negation of the Continuum Hypothesis as an axiom and get a different set theory. A similar thing was going on with the fifth postulate. We got two geometries and two set theories, each of which was as consistent as the other. This was a big blow. How could one have certainty when there appeared to be two consistent theories saying different things? So, we got to my last line of defense. I acknowledged that there are two logically consistent geometries but argued, like your grandfather, Ravi, that only Euclid's version provided truth about the universe, because only his axioms were true. And now even this last hope may have been proved false. Human beings can never be certain about anything, and without certainty there cannot be meaning."

I wanted to rebut that somehow, but couldn't.

• • •

That night I found the story of Bauji's release in *The Morisette Chronicle:*

SAHNI RELEASED FROM PRISON!

MORISETTE — In a surprise culmination to a case already marked by numerous twists and turns, Governor Williams today ordered the release of Mr. Vijay Sahni from prison. The Governor did so after Judge Taylor recommended that the state should not press blasphemy charges. Both Mr. Sahni and Judge Taylor have declined to comment on the fresh development but intense speculation now surrounds the conversations the two are reported to have conducted in prison. Sources have indicated that mathematics was indeed the subject of most of the exchanges between the two, but experts remain unsure about the relevance of this to the legal issues that may have come up before the judge.

The developments today have come as a complete surprise because it was believed that Judge Taylor had made up his mind in favor of pressing charges. Informed sources tell *The Morisette Chronicle* that several weeks ago Judge Taylor had actually drafted a letter to Governor Williams recommending that this case go to trial. It was said that while the judge was sympathetic to some of the points made by Mr. Sahni, he still felt that on the weight of evidence available, he had no choice but to recommend that the state proceed with the blasphemy charges.

Judge Taylor today restricted himself to stating that he had submitted his views in a detailed letter to the governor. Sources have indicated that the governor is unlikely to make the details public in the highly charged atmosphere surrounding the case. The governor had found himself caught between the criticism emerging from the

New York liberals who had made this a test case for the right to freedom of expression and his own conservative constituency's vocal demand that Mr. Sahni be punished for deliberately affronting the Christian sentiments of our community.

The governor, while announcing Mr. Sahni's release today, chose to placate this constituency by stating that while his personal views on the matter were well known he could not contest the recommendation of a legal expert of Judge Taylor's standing. His decision was made known shortly after noon today, and Mr. Vijay Sahni was released in the evening after the requisite formalities were completed. He refused to talk to journalists, but did stop briefly to acknowledge a few acquaintances among the curious crowd that had gathered outside the courthouse in anticipation of his release. There were a few hostile voices condemning the release, whom Mr. Sahni ignored.

While legal experts are unclear about the reasoning behind Judge Taylor's views, most do believe that, given his past reputation, his stand was likely based on sound legal thinking. They further indicated that although the decision to not press charges means that in the strict sense this case does not become a legal benchmark for proceedings based on the controversial blasphemy clause, it does, however, make it practically impossible for the law to be applied in the future.

Accompanying the article was Bauji's picture. The caption under the photograph announced that he was boarding a train for New York, from where he was scheduled to embark on a ship bound for London. Bauji stood at the door of the train car, dressed in a traditional Indian kurta-pajama, facing the people of Morisette one last time. In front of him was a group of what appeared to be reporters, several of whom wore faces of apparent dismay and disapproval. Behind them was a line of constables holding a throng at bay. Many in the crowd waved placards whose writing I could not make out in the photograph, but their sentiments were clear enough. Bauji appeared not to notice them at all. He looked instead at one tall man standing slightly apart from the half-circle of journalists. It took me a few seconds to be sure that the other man was indeed Judge John Taylor. The judge, like Bauji, ignored the din around him. He looked squarely back at my grandfather. I realized, with a start, that both men were unquestionably happy. The angst that was so palpable in the last transcript was gone: Bauji had the tiniest of smiles on the corners of his lips, and he and the judge appeared to be sharing some private amusement. They had figured something out together and now possessed the camaraderie of joint discovery.

What had turned it around?

• • •

The day after reading about Bauji's departure to New York, I left for that city myself. I was headed for my final round of interviews with Goldman Sachs. On the plane I tried to prepare for the interviews by brushing up on my finance lecture notes, but every time I'd read a paragraph or two, an extraneous thought would intervene, and by the time I would sort it out and get back to the text, I found that I needed to begin at the beginning once again. Finally I fell asleep, which under the circumstances was probably the best thing.

From the minute I landed in New York, I was quite taken by the Goldman panache. There was a car to pick me up at the airport, the hotel was plush—obviously expensive, but not ostentatious—and there was a welcome packet waiting in the room inviting me to use room service and order a meal to "renew and refresh" myself from the long journey. Wouldn't take much to get used to this, I thought. Of course, I did need to get the job first.

I hit the bed expecting to toss and turn with nervess, but it didn't happen. Instead, the softness of the sheets and the perfect firmness of the mattress conspired to induce a quick sleep. My 7:00 a.m. wake-up call interrupted a dream in which I was trying on a golden bracelet that, according to Peter, was the latest fashion accessory amongst those in the know.

The interview questions I faced at the Goldman offices were rigorous and well thought out. Most of them were about valuing companies or the use of debt instruments and the like, but one of my five interviewers (Harold Smith IV, no less), asked me, of all things, a mathematical problem of sorts. "It helps me pick out the guys who can think," he said. Here's how he asked the question: "Do you agree or disagree that at any given time in New York there live at least two people with exactly the same number of individual hairs?"

He can't be serious, was my first thought. But his interested, expectant half-smile said otherwise. I'd better think about this, I told myself.

But what an odd question! Is there a pair of people in this city who have exactly the same number of hairs?

"Do you mean on the whole body, or just the head," I asked.

"The whole body," he said.

I dimly recalled hearing about a condition which renders some people completely hairless. It seemed likely, or at least possible, that at least two people in New York City had this condition (in which case they would have the same number of hairs, i.e., 0).

But my interviewer didn't bite on that line of reasoning. "Let us agree to set that possibility aside."

Hmm . . ., I wonder how many hairs I had on my body. That would depend on the average number of hair per square inch. I looked at my forearm and imagined a 1 inch square; there looked to be well under 50 hairs in my imaginary square. Clearly, though, the density of the hair on my head would be much higher. Perhaps that was a piece of information I could inquire about.

"Roughly a thousand per square inch on a nonbalding head," said Harold Smith IV. His smile told me that I was likely onto a promising line of thought.

Approximating the human body to a box, I calculated that the hairiest of all possible people had no more than 7 million hairs on his body (the pronoun gender choice in this case being deliberate). And this person would be about as hairy as a gorilla.

"What's the population of New York?" I asked.

"About 9 million," said my interviewer, now knowing that I had cracked the case.

So there it was. Every person in the city had a "total hair number" ranging from 1 to 7 million. But since there were 9 million people, there had to be some duplication in hair numbers.

"Bravo!" said Harold Smith, clapping soundlessly.

I found out later that his problem is an illustration of the pigeon hole principle, which states that if n pigeons are put into m pigeonholes and n is greater than m, then there's a hole with more than one pigeon. It's an obvious enough statement but it turns out to be useful in many dozens of proofs.

After filling out some forms I was politely thanked for my interest in Goldman Sachs. "We decide very quickly," said Stella Channing, the HR manager. "You should hear from us tonight, or at the latest by tomorrow afternoon."

All in all a good day, I thought, joining the crowd of power-walking professionals streaming down Manhattan's Broad Street. They were just getting out of work and still seemed purposeful and intense. They had the air of people who had worked hard all day and were now going to enjoy their friends in some cozy corner of the city. I could see myself here, in this city, being with these people. "God, I hope I get the offer."

It turned out I didn't have to wait very long to find out. Just before I got on the plane, I checked the messages on my answering machine. Sure enough there was one from Ms. Channing. "Ravi, I am very pleased to tell you that Goldman Sachs will be extending you an offer. You did very well today and were strongly supported by everyone who interviewed you. We will be confirming our offer in writing, by the end of the week."

Twenty seconds later I was talking to my parents in India. Not unexpectedly, they were jubilant.

"I've imagined this moment for the last ten years," said my mother.

• • •

When I got home, just for the pleasure of it, I hit "Play" on the answering machine. As before, Ms. Channing's voice was cordial, and I could hear her smiling as she told me once, twice, and then a third time, that I had indeed hit it out of the park. My happiness, unlike the excited thrill when I first heard the news, had settled down to a steady state of slow, deep contentment, and it wasn't until the next morning that I noticed that there was another message on the machine.

"Ravi, it's Nico. I hear from Claire that you're seriously considering becoming a banker of some kind! That would be a complete and utter waste of your life. You have the talent and the genes to be a fine mathematician. I am arranging to get you a full scholarship for graduate work in mathematics starting in the fall. Come by my office tomorrow and I'll give you the details."

Wow. Here, then, was a choice: Goldman offered the end of worrying about money, the happiness of my parents, a new city, and a life of doing, not just observing. Graduate school meant pursuing a subject that delighted me and, perhaps, the possibility of Claire. How was I supposed to decide this? Moreover, how could I possibly be certain that I was choosing the right path?

When I called Claire to find out what she knew about Nico's offer, she asked me instead if I had seen the package she had left by the front door of my apartment. "It's several pages of Judge Taylor's personal notes that I got from the National Archives. My mom thought that it was a good bet that his personal papers would be there, and I've spent my last two Sundays looking. I think you need to read them."

CHAPTER

1 2 3 4 5 6 7 8

JOHN TAYLOR FEBRUARY 1930

Four days ago I set sail from New York City to London en route to Bombay. I am going to see my friend, Vijay Sahni. It has been over ten years since we met and while we have periodically written to each other, I must confess to a feeling of childish excitement in anticipation of seeing him once again—a feeling I'm surprised still lives inside me.

Marcus Aurelius, the Roman Emperor (and a fine philosopher, as well), once said: "Severally on the occasion of everything that thou doest, pause and ask thyself if death is a dreadful thing because it deprives thee of this."

My long-ago jailhouse conversations with Vijay easily surpass Aurelius' challenge. I frequently find myself replaying snippets of these talks, even capturing his exact tone and pitch in my mind. His words fill out the vaults of my memory, quite disproportionate to the fleeting hours within which they actually unfolded. The first proof he showed me was the Pythagorean theorem, and I have visualized its construction so many times that I reckon I could recreate it in my sleep. His words accompanying the proof were sparse, but precise. He seemed to know where I would get stuck even before I got stuck. More importantly, he knew exactly what to say to get me out of my logjams. He had an instinct for teaching.

Nothing remotely like that has been available to me since, and that is especially true of what passes for discussion on this ship. My only human interaction consists of listening to Mrs. Merriwether's litany of health complaints over breakfast and dinner. She talks, and I pretend to listen and hope that the timing of my nods matches the spots in which she expects some acknowledgement from me. But most of the day is my own and I've taken to going to the lower deck and looking at the ocean and writing in this journal, undisturbed by the other passengers.

When I first saw him I was shocked by how young he was. Given the consternation he had caused, I had expected to find an older man, more rigid in his beliefs and more practiced in his eloquence. But when I saw him I realized that he couldn't be much older than John Jr., except John Jr. was finishing his last semester at Princeton College, while Vijay Sahni was prisoner number 1729 in the Morisette County Prison. So my first instinct towards him was paternal. He looked too skinny and too weak to be dangerous to anyone. Between his recently grown beard and the curly black mop on his head, his face was barely visible. I wanted to give him an out, to find a reason to free him and get the whole mess over with. I wanted him to say that he had uttered his blasphemy in a fit of emotion and that he meant no harm. But he did not cooperate: "You have it wrong, Judge Taylor," he told me, his chin jutting out.

As a rule I was (and remain) very patient in all my dealings. But Vijay seemed too sure of himself, too cocky, and when he told me that his mathematical analysis had helped him conclude that God had come up short, I could not suppress my irritation from surfacing. I remember being quite taken aback by the ferocity and extent of my anger. It was the first time that I consciously realized that I had a strong emotional attachment to my faith, and this young man was treading on soft territory, too close to my core. My faith was—and is—the fundamental bedrock of how I see the world and being exposed to a world view which had no need for it was deeply frightening.

From the beginning, though, Vijay had one redeeming quality: he could do mathematics. So I listened to what he had to say.

When he told me about the axiomatic way, it seemed to me to be the most natural thing in the world. You had to start with something, so you started with statements that you had no doubts about. Things like, "It is possible to draw a line from any point to any point." Who could argue with that? And there was so much power to be had from assuming a few simple axioms it seemed impossible to even consider that the axiomatic method was somehow flawed.

My argument was not with the method, but with the content of the axioms. I wanted to include one of mine in our discussions. Who wouldn't intuitively accept that "everything is created by something" is an obvious axiom—it's just a cousin of the "first cause" principle. But Vijay had been stubborn in his opposition. What annoyed me—no, exasperated me—was that he was unable to provide a satisfactory reason for rejecting my axiom. It seemed to me that he was rejecting my axiom because he did not like where it led. I genuinely believed that my axiom was just as certain as any of his were, including that accursed fifth postulate. But he didn't budge, and as an agent of the law, I was obliged to do my duty.

Thinking of the night when I decided to let his case go to the jury still makes my jaw clench. I knew that I was in essence letting Vijay go to jail. The rub was that by then I had begun to understand him, and though I wouldn't have admitted this to myself, I had also begun to admire him. But nothing mattered other than the law. It was my duty to uphold the law, and the plain fact was that Vijay's methods weren't the problem, Vijay's rigid beliefs were the problem. Why else would he have rejected my axiom?

And then everything changed. A newspaper headline reported in essence that Euclid's fifth postulate was false. I remember reading the article several times trying to make sure that there wasn't another explanation, some heretofore unseen way to keep the fifth postulate and Eddington's observations in coexisting harmony. But nothing budged. I kept telling myself to remain calm, but something stark must have shown on my face, for Mr. Hanks asked me if I was all right with more than a touch of worry in his voice. "Fine," I had replied, as I headed to show the article to the one man I could talk to about this.

On the way, I kept trying to suppress the rising sense of panic that had taken hold of me. "Why," I asked myself, "should I let these arcane astronomic observations from some distant land affect me in this manner? How did they change anything that was important to me?" But despite my sensible attempts at denial, the knot at the pit of my stomach kept getting tighter.

Vijay's reaction had turned out to be quite different from mine. He read the article carefully once, then again, and then one more time. He closed his eyes awhile and then in a moment that I can still see so clearly in my mind's eye, his shoulders slumped and he nodded his head, calmly accepting the cold fact that those two columns of newsprint had shattered the possibility of absolute certainty, an idea around which he had based his life. He didn't fight or deny or look for excuses. He just acknowledged that he had been wrong. Only then did I gather the courage to admit the source of dread to myself: if Euclid's fifth postulate could be false, then anything could be false, including my blessed faith.

That night I let myself see the world as an atheist must: a desolate planet occupied by people who had abandoned themselves to amoral meaninglessness. At least Vijay's atheism had a thirsting for knowledge, but if his methods were unreliable at their core and if the Good Book was mere fanciful allegory, then there really was nothing left. There could be no hope, or reason, to understand anything. The universe would be nothing more than an accidental pile of rocks, revolving around each other, without purpose, and all human life would be a tiny, utterly inconsequential speck of nothing, existing for a brief moment in time, in a far corner of a below-average galaxy barely causing a wisp of a stir, not remembered or even noticed.

It was this Ecclesiastian view that grew on me the night of the Eddington announcement. Doubt, like a ravenous cancer, settled into every one of my memories. Everything in my life, even what I considered to be my best moments, seemed to me to be mere vanity. The first in my family to finish college: vanity. First in my class at Princeton: vanity. Even my (now failed) quest for a seat in the Supreme Court: vanity. For the first time in my life I understood what Koheleth had been saying in The Old Testament:

> *Vanity of vanities; all is vanity . . . What profit hath a man of all his labor which he taketh under the sun? One generation passeth away, and another generation cometh . . . The thing that hath been, it is that what shall be . . . and there is no new thing under the sun . . . When I applied mine heart to know wisdom, and to know the business that is done upon the earth . . . then I beheld that a man cannot find out the work that is done under the sun . . . Yea further; though a wise man think to know it,* yet shall he not be able to find it . . . *That which is far off, and exceeding deep, who can find it out?*

Koheleth wrote as if he had glimpsed the impossibility of Euclid's system, and hence the impossibility of finding absolute truth and lasting meaning.

The hour got later and later but sleep remained impossible. I paced in my room for what could have been a few minutes or a few hours. Thoughts arose in parallel and jostled against each other for prominence. My back hurt. Memories from my childhood interspersed with sudden bursts of unfocused anxiety swirled in my mind without conclusion or resolution. Seeking to break the spell, or perhaps merely to increase the amplitude of my pacing, I headed out to the yard. But once outside, I felt a strong desire to walk without retracing my steps, and somewhat to my own surprise, I found myself walking out onto Church Street. I moved fast, without destination. Speed spoke of action and action quelled worry. I told myself that I was being childishly impulsive; walking around aimlessly in the dark would not yield any answers. But it felt good to walk and turning back would have felt like surrender, so I pushed forward. I must have walked further than I realized because I remember noticing at one point that there were no street lamps, which meant I was outside Morisette County. The homes had grown sparse, fewer and further apart; I couldn't tell where I was anymore. There was no one to ask for directions. This was just as well, for had someone chanced upon me I would have cut a strange sight indeed: a disheveled, solitary figure in the dead of night, with a vintage army winter coat hurriedly thrown over wrinkled trousers (never before and never since have I ventured out in that state). "I'll see where I am at light," I told myself and walked on.

At some point on my march it occurred to me that without explicitly deciding to do so I had agreed to frame the discussion of my beliefs on Vijay's terms, not mine. The thought came to me when I imagined what my father would have made of this business of the fifth postulate turning out to be false. I could see him brushing aside the question with one shake of his head. They call it faith because it requires you to accept, not question. Deduction and argumentation is too superficial, he would have said. The very question of the fifth postulate would have been utterly meaningless to him, for he would have rejected its terms. Yet such was the power of the axiomatic method—with its beguiling promise of absolute certainty—that I had allowed myself to be seduced into Vijay's world. And now I was as trapped as he was.

Just before dawn I saw some lights in the distance in the shallow valley below me. As I got closer I could make out individual buildings, and from the silhouette of the short and somewhat stout steeple, I realized that I had come upon Dogstown, a predominantly Negro hamlet whose old mill had been converted into the village church. They had run out of money halfway into the construction and thus had to curtail their ambitions of building a fifty-foot steeple, which would have equaled the tallest one in Morisette. Instead they ended up with a curious, wide-based, fifteen footer, which in its own earthy way was not unpleasing to look at. Dogstown meant that I had walked a little over eleven miles, or perhaps more, depending on if I had happened to take the Greenwood turn. Despite the cool night air I was quite thirsty. The backache which had subsided with the fast trek had reappeared, and moreover my feet, trapped in my work shoes, were beginning to blister at the soles. I finally made it to the church, the last quarter of a mile requiring more effort than the previous ten had.

The church was locked and there was nobody around. There seemed to be a dim light coming through the opening under the main door but I wasn't sure if it was from a lamp or from some trick reflection of the rising sun. Not knowing what else to do, and too tired to consider my options, I sat down, resting my back on the wall next to the door.

Next thing I knew, someone was asking me to wake up. "What are you doing here, sir? Wake up! Wake up!" I emerged from sleep unsure if I had slept for an hour or a week. "Where am I?" I remember asking.

"Why, you're in Dogstown, sir. What are you doing here?"

More of an accident than anything else, I said. The man inferred that I had been in an automobile mishap, and not knowing how I would explain my purposeless wandering, I didn't try to correct him. "Yes, yes those automobiles are more trouble than they are worth," he said. "Come on in, you sure look like you could use some food and drink." My host turned out to be Pastor Darrel Huston. His room at the back of the church was spare but clean, and he offered his only chair to me. "Take a load off," he said. After I had gulped down two tall glasses of water, he offered me some warm homemade bread with butter and fresh plum preserves. To this day, that humble repast is one of the most satisfying food memories I have. It easily beats all the culinary excesses served nightly on this ship.

After Pastor Huston judged that I was sufficiently recovered, he invited me to attend the Sunday service. I asked him what time the service was scheduled to commence, for at Morisette we always started at 8:00 a.m., regardless of the weather, and the hour seemed a lot later than that. "We begin when everyone gets here," he said. And did he have his sermon ready? He laughed, "We don't do sermons here, Mister; we sing."

I was the only white person at the Dogstown Church that morning. In fact, I was told that I was the only white person ever to attend services there. People were openly curious, and I must have fielded a dozen separate inquiries about who I was and what I was doing in Dogstown. Finally Pastor Huston made an announcement welcoming the "esteemed John Taylor" from Morisette County who had had car trouble and then got lost in the darkness.

"John Taylor? Ain't he the judge?" asked a voice from the back. "Yeah, yeah! He's the judge doing that guy's trial." But before there could be a fresh round of questioning about Vijay, Darrel signaled the organ player to begin and the Sunday service was officially underway.

I've spent many sublime hours in churches in my life, but that shining afternoon stands out from the rest: it was more personal, more intense, more emotional, and in the end more meaningful than what I had been accustomed to, or what I had any right to expect.

Pastor Huston began with a brief sermon. His congregation was active and vocal, and frequently backed him up with loud comments such as

"Amen," "You said it now, Brother," and "Preach it now." The Pastor's delivery was very rhythmic—more like a song than a speech. He got more and more inspired and devotional as the loud contributions from the congregation increased.

After that everybody sang. Every last person. They all began together, reading from well-worn sheets of music. But then Pastor Huston would call on an individual member, say "Fat Aunt Sally at the back" or "Jewel Thief Robinson in the corner," and he or she would sing solo, often making the song his or her own. They would improvise new words and fit them into the song's chord progression. Many of the congregants had their own musical instruments, such as drums and horns, which they would bring out when it was their turn to "testify." The rest of the group would listen and clap, and sometimes dance in the pews. The soloists, too, would often do a dance of their own which, while graceful, was inappropriate—or so I thought at first blush—in a place of worship. Thank the Lord, I was not invited to participate.

When the session started I thought that it more resembled a boogie woogie saloon than it did Sunday services. Our church in Morisette would begin with a Bible reading, followed by a sermon, followed by silent prayer. Silence was in short supply in the Dogstown church. I wondered how anyone could focus on anything, let alone the Lord, in all the din.

The faces around me seemed to tell a different story. Despite everything, there appeared to be purpose here. At first, cynical as I had become in my time with Vijay, I thought that the parishioners were just putting on a good show; perhaps they were trying to outdo each other with displays of genuine devotion. But this was no display. Something very real was going on in the Dogstown church, and I didn't quite understand it.

In time, I grew tired of my thoughts and began listening to the music. The songs were mostly unfamiliar to me. I know a little something about music and I could tell that many of them seemed to be based upon the pentatonic scale, which is a common scale in African music. As its name suggests, the pentatonic scale is a five-tone scale, such as that produced by the five black keys of the piano in succession: F#-G#-A#-C#-D#. Some of the younger singers abandoned their traditional sound in favor of the new-fangled blues scale, which to my surprise I quite enjoyed.

Then the songs grew older, slower, and somehow sadder and more soulful. I began to recognize spirituals that my Negro nanny would sing to me as she would tuck me into bed when I was six or seven. That long-ago bridge was opened to me by the unbearable loveliness of the music. A feeling of belonging mystically gained entry to my heart in the notes of that music. Promises of serenity. Guarantees of strength. Songs of love. Peacefulness washed over the faces of the congregation along with the sun shining through the stained-glass windows. They had done what they came there to do. It was suddenly so clear that a common thread runs through all things, living and dead. I felt connected, connected to the congregation, connected to Dogstown and to Morisette, and connected to the universe—even the non-Euclidean universe.

Amidst this onslaught of grace, I was quite aware that I would never be able to understand it, to say nothing of proving it. I could analyze the music, analyze the feelings of brotherly love, perhaps even understand my emotions, but I would never be able to understand this grandeur of the connectedness of all things. My doubt from the night before was gone. I found myself crying.

Later that day Pastor Huston arranged for two horses and some food and water for my return trip to Morisette. "I'll ride with you so I can get Blackie back home," he said. Blackie, a spotless white stallion, went easily under saddle and turned out to have a fine gait.

Before we left Dogstown, I offered Darrel some money for everything he had done for me, but he vigorously shook his head as soon as he saw me pull out my wallet. "It would have been unthinkable for me not to come to the aid of a tired and hungry traveler at my church's doorstep. I do not aim to profit from merely doing my duty," he said in a tone which didn't encourage further discussion on the topic.

The road to Morisette was wide enough for us to ride alongside each other, and neither of us was in any hurry. We rode easily and talked easily, moving from speculation about how the newly formed Negro League teams might do against the Yankees (we both thought that Oscar Charleston from the Negro team in Indiana could be the starting center-fielder for the Yankees), to our fathers (both of whom were preachers), to the prospects of the Morisette paper mill, which employed most of Dogstown's men (the new presses would surely eliminate jobs).

Halfway to Morisette we stopped under an old oak tree and had some water and a snack of green apples and honey. Perhaps sensing that I wanted to talk about this morning, Darrel asked me what I thought about the service.

I paused awhile before answering. I wanted to get the words right and do justice to the depth of my feeling. But nothing really came. "It was not what I expected," was all that I could muster up in the end.

"All the singing must have been alien for you?" he asked gently.

"Singing is not a form of prayer we white folk are accustomed to."

Darrel nodded his head. "It is just our way and it seems to work," he said.

We rode in silence for a while, passing the Greenwood farm and orchards. Then, perhaps because I was in a different state of mind, or perhaps because I felt I could trust him, I asked Darrel a question that I couldn't have imagined even formulating a week ago. "Pastor, do you ever doubt?"

"Doubt what?" He was immediately interested.

I didn't hold back. "Doubt your faith; worry that perhaps God does not exist? After all, there is no proof."

Unexpectedly, he laughed. "There can be no proof. The acceptance of God can only come from faith. Faith is a starting point and you can never prove a starting point because, well, it's a starting point!"

"Like an axiom," I said.

"What?"

"Nothing."

Darrel squinted his eyes at me. He took my question to indicate a crisis of faith—which in some sense it was—even though the wonderful ceremony he himself had orchestrated had given me back much of my balance. But what he said to address my crisis was not at all what I would have expected.

"Judge, it is natural to have doubt. But you must remember that much more important than one's starting point is what one does with it."

"What do you mean?" I asked.

"I'll take the example of my own life. I am sure that God exists; I accept this on faith and I don't need to find a proof for it. The rest of my life flows from it. How I work, how I interact with my family, my parish, and my community, what I do with my spare time, how I pick right from wrong, all of this derives from my belief in God, and it is in this sense that my belief is a starting point for me. But the starting point would be worthless if I did nothing with it. One honors God not by merely believing in him, but by living one's life to embody what he stands for."

"Yes, but what about the starting point itself? What if the starting point itself if false?" I asked.

The utter rejection of this premise that I was expecting didn't materialize. "Sure it's possible that my starting point is false," he allowed. "I cannot believe that something so beautiful can be untrue, but this is not something I have proof for. Quite frankly, different people come up with different starting points according to the nature of their souls or the sum of their experiences. In Dogstown the starting point of our worship includes music, and yours does not. But really the particular starting point does not matter as much as we might sometimes think. Grace can happen from a variety of starting points."

"If our starting points are dictated by the particular nature of our souls, then it would seem to me that one starting point is no truer than another?" I asked the question I thought Vijay would have asked.

He smiled and shook his head. "Judge, I think it's important for a man to have some kind of starting point that can act as a unifying principle for his life. As long as he is true to some core beliefs, he can't go too far wrong. Which starting point is true is not something we humans can make much progress on."

When I got back to Morisette, I had expected to find Vijay still reeling from the stunning implications of the Eddington announcement. Instead, for the first time ever, he greeted me with a wide smile.

"You're happy?" I asked, on the verge of disbelief.

As it turned out, he was indeed happy. That morning, he had received a letter from an Indian mathematician named Ramanujan. Vijay had briefly known Ramanujan in India and then had met him at Cambridge University. I later learned that Ramanujan was one of India's greatest mathematical geniuses. Indeed, he was one of the greatest mathematicians the world has ever seen.

"Judge, I'm really thrilled," said Vijay. "Ramanujan has sent me a letter with some truly fantastic theorems. Let me show you." He handed me two densely written sheets of paper from a heavily stamped envelope.

At this point I didn't know anything about Ramanujan and I didn't really know what to look for. His handwritten equations seemed strange, filled with square roots and fractions that seemed to extend forever.

Vijay, as he was wont to do, was itching to teach. "Let me show you some of these. They are truly remarkable, even awesome!"

"Wait," I said. Something in my tone must have caught his attention, for he stopped looking for his pen and notebook. "What's wrong?"

"Nothing," I replied. "Sit down." I had his attention. He sat still, looking at me. "When I last left you," I said, "we had concluded that mathematics could not provide certainty about our world. And here you are today, captivated by some new set of equations. Has something changed?"

For the first time in our conversations, Vijay didn't seem to have an answer at the ready. When he spoke, it was in a quieter, almost humble voice. "You are right. If mathematical proof is suspect, then these equations are suspect."

And then after a pause he said, "But the thing is, they seem so utterly true." He let out a long sigh. "I was really distressed over the last two days. Everything I had held true for so long had been disproved, and I didn't quite know how to deal with it. The very sight of food would make me nauseous, and I couldn't sleep at all. I wanted to walk, but of course that is not possible here. It was a terrible state that I can't fully describe in words. You would have to experience it to understand."

But of course I had experienced it—and that too last night. "So what happened?" I asked.

"This morning Ramanujan's letter arrived and nothing else mattered. The equations had me entranced and I had not even given a thought to what they actually mean in the context of absolute certainty. They are immediate and real; I couldn't doubt them even if I wanted to. But at the same time I know that I can never be absolutely certain of them, and in a way that doesn't even seem that important. Working with them is what is important. I don't really understand what is happening."

It seemed to me that Vijay was working from a starting point that somehow guaranteed that his mathematics was real and true. And his starting point had nothing to do with the axioms we had been working with. They were a front for something else and I needed to find out what this something else was. "I think I've had some realizations that may help us make some progress on this apparent paradox," I said. Vijay sat back on his chair, listening. I sensed that our roles were reversed, and that this time I was the one leading us forward. Now, many years later, in trying to recall how the conversation went, I realize that I modeled my argument after Vijay's didactic style.

"Vijay, you've shown me lots of theorems, all of which are true if you believe Euclid's axioms to be true. But we have seen that these axioms are not true of the space around us. So what does that say about the truth of these theorems? Specifically, is the Pythagorean theorem true?"

He pursed his lips. This was a difficult topic for him to talk about, just as his questioning of the truth of religion, and Christianity in particular, had been difficult for me. But Vijay was a truth seeker above all and he called it the way he saw it. "The Pythagorean theorem is conditionally true. If Euclid's axioms are true, as they are on a plain sheet of paper, then the Pythagorean theorem is true in that context, too."

"Exactly," I said. "And these theorems of Ramanujan, they are likewise conditionally true, for presumably they depend on another set of axioms?" Vijay thought about that for a long time. In the end, he allowed that just as there were axioms underlying geometry (both Euclidean and non-Euclidean), there were other axioms underlying number theory. And because it seemed that even the most apparent axioms are impossible to verify with certainty, the theorems of Ramanujan may also be conditionally true.

It was time for my inductive leap. "Looks to me like all of mathematics is conditional. It is based on one set of axioms or the other." Vijay nodded his head, very reluctantly, but he did nod. "So mathematics, then, is like a game—somewhat like chess," I said. "You have some starting conventions, or axioms, and then some rules of inference, and you take the axioms that may or may not be true into theorems that are conditionally true. This game does not mean anything. It is just a game."

I had reached the limit. Vijay was vigorously shaking his head, disagreeing. "Judge, I can't disprove what you say, but I know from experience that your statements do not capture the spirit of mathematics. It is true that all of mathematics, be it geometry, or number theory, or even this new subject they are calling set theory, all of it is based on axioms. But the subjects do not exist because of their axioms; quite the reverse! I deeply believe that mathematics is not a mere chess game where arbitrary axioms devised by humans lead to arbitrary theorems. No, I do believe that mathematical truths have an external reality quite independent of the minds considering them. Indeed, I think that mathematical truth exists outside the matter of the universe and independent of any consciousness. This truth is timeless and absolute. It as at least as real as a stone or a tree or this prison cell. There would be no point in doing mathematics if you didn't believe in this external reality of mathematics." He was silent for a minute and then said almost to himself, "I cannot believe that something so beautiful is devoid of truth."

He didn't know it, but he had echoed Darrel Huston from a few hours ago. I had half expected, or at least hoped, that Vijay would say something close to this. I was ready for him and I leapt at the opportunity. "You say there is this external reality of mathematics. Well, where is it? Is it out in space somewhere? Written in a big book of equations, perhaps? You say you know this structure exists, but you also tell me that it can't be verified, because the axioms you use to discover it can't be verified. This structure, then, is something you are asking me to take on faith."

I noticed that my voice was getting louder as I spoke; 'faith' came out at a near shout and it occupied the cell even after I was silent. Vijay averted his eyes. After a minute or so he started to say something, then stopped himself and leaned back. Finally he spoke without looking at me. "So you're saying that it is like your faith; that my belief in the external reality of mathematics is akin to your belief in God?"

I needed to tell him about Dogstown. "Yesterday I happened to attend a church service where the congregation sang in prayer. This singing was completely new to me. I pray in silence or by reading the Bible. But these people sang. They sang with great emotion. Contrary to all my expectations, I was greatly moved, and later the church's pastor made me see that it is impossible to start anything without faith in something. The only reason you do mathematics is because you believe that mathematical objects exists outside the human mind. It is your starting point. Just like my starting point is a belief in God. But both of our starting points are, and must remain, articles of faith."

Vijay looked at me but he had a faraway look in his eyes. Then he rubbed his hand on his chin, looked at Ramanujan's equations, then back at the drawing and slowly nodded. "I think you might be right." I thought he might say more, but no, that was all. Agreement.

Our accord was so swift and sudden that it took me by surprise. I had expected him to resist this equivalence vigorously; instead, it seemed apparent that he had been thinking along similar lines. Only then did the magnitude of what I had done strike me. With one fell swoop I had equated my faith in God with a mathematician's belief in absolute mathematics, both of which were just beyond the reach of our reason. I got scared and anxious as if I myself had committed some blasphemy. But I saw that I was not alone, for Vijay too was confronting the loss of his universality and absoluteness.

He said, "What is quite amazing is that we humans gravitate to such different things. That our . . . faith . . . takes such different forms."

It is, indeed, amazing. The human experience is such that we yearn to find something lasting and true, something that speaks to our own hearts and has meaning. Meaning, however, whatever its variety, seems to demand faith.

The next day, I sent a note to Governor Williams recommending that Vijay Sahni be released. The governor called me upon receiving my message. He wanted to hear my reasoning on the matter. "I think he sees things a little differently now," I said, neglecting to mention that I too, in at least equal part, saw things differently myself (although all that did come out later).

The very next day Vijay was released. He was to board a train to New York City to commence the same voyage that I currently find myself on. He has since gone on to become a mathematics professor at Delhi University. We have corresponded frequently over the years, mostly about mathematics. My interest in mathematics has not subsided, although it has changed form, from geometry to number theory, in part inspired by the dazzling results of the late Ramanujan.

I never did read the letter Ramanujan sent to Vijay that day. I will see if he still has it. Meanwhile I can barely wait to be in Bombay.

[Continuing several days later] Even as the Queen Victoria docked in Bombay, I could see a large horde of people converging toward the unloading

area. They apparently all must have had business there, but I could scarcely imagine what activity could occupy such large numbers. I reckoned that the one and a half city blocks around the ship contained more people than the entire population of Morisette. And what colors they wore! The women, who were no less aggressive than the men in jockeying for position, had on saris of bright reds and yellows. The men wore white, making their brownness more visible and, I thought, more becoming.

Once I got down the ship's walkway, closer to the crowd, I became aware of a pervasive aroma I couldn't quite place. A few day's later when I asked Vijay about it, he identified its key constituents: incense and sandalwood from the sadhus, dung from the cows that seemed to be on every street corner, and spices cooked in oil emanating from improvised kitchens on carts that served up some truly delightful delicacies. "It is the smell of India," he said.

I had anticipated finding Vijay in the passenger receiving area, but it turned out that only the British were allowed access to the incoming passengers. I was greeted by a small entourage led by a British army man, Colonel something-or-other. He had been notified to expect "an American judge," but the nature of my business in Bombay had not been communicated to him. When I told him I was in India to visit a Mr. Vijay Sahni, he didn't quite know what to make of it. "Please allow me to make some enquiries," he said, raising his exceedingly bushy eyebrows.

His enquiries mustn't have been fruitful, for an hour went by without any news of Vijay's whereabouts. I passed the time by observing the crowd below—day laborers hoping to make a few paisas by unloading the ship's cargo. But there was far less work than there were people. Only the ones who had pushed themselves close to the hull were fortunate enough to get work, and presumably, get paid. Our economic problems in Morisette were nothing compared to what was going on here.

My observations were interrupted by some commotion immediately outside my waiting area. An Indian man appeared to have taken on the two British constables who were preventing him from entering. "No Indians allowed in here, man. Whites only. Can't you read the sign?" The man said something back to them and while I couldn't make out the words, I could identify the voice.

It took ten minutes of explaining to get the constables to allow Vijay through. "Bloody British," he said, finally in, shaking my hand.

He looked older that I expected. His eyes seemed to have sunk in a bit and I saw thin lines crisscrossing his face. He was stronger than I remembered him. His body had a tightness it didn't have in Morisette. "Let's go," I said.

But that was easier said than done. The Colonel couldn't believe that I wanted to venture out with an Indian man without any special travel arrangements. "The food will make you sick and the mosquitoes will drive you crazy," he warned.

But making special arrangements would have almost surely excluded Vijay, and that wouldn't have been right. Despite all the class privileges the British wanted to award me—a fellow white man—I was here to see my friend, and one does not disrespect one's friends. It is not the American way.

So off we went, he and I, to the Bombay Central Station. Vijay went to get the train tickets and I became aware of being stared at without respite. In America, if you catch someone looking at you, he or she will briefly nod and then look away. In India, especially if you happen to be white, it's different. People stop what they're doing, and gawk. "Just ignore them," said Vijay, "Stare back, as a matter of fact!"

While we waited for our train, I had my first Indian meal consisting of a flat bread and some black lentils. I thought it was delicious. My stomach, however, didn't see it quite the same way. I got ill two hours later and stayed ill for the 3 days it took us to get to Delhi. Probably dysentery. But Vijay was a lifesaver. He applied cold compresses to my forehead and was somehow able to procure a continuous supply of rice and yoghurt which were the only things I could keep down. In my stronger moments, I tried to talk to him—I had come this far for that very purpose—but he shook me off. "You need to rest," he would say, and he was right. I slept the last 18 hours of the journey, and by the time I woke up, I felt that I would perhaps survive.

His house in Delhi was small, but it had a big yard around it. There was an old banyan tree in the back and two mango trees, which made a shady canopy in the front yard. Inside, he had built bookshelves on all of the available walls but there were still more books than there was space. Most of his books were on mathematics, but there were exceptions: a treatise on the birds of North India, a well-worn copy on gardening, and of all things, a copy of the American Constitution. There were no pictures or art, save for one photograph of a young woman. After much prodding I found out (the next day) that she was Vijay's fiancée.

So he would one day be married and have a family. I wondered if his children would be mathematicians, or if they would, as children often do, rebel against their father's life and go out and create their own destinies. Perhaps, the grandchildren, then . . . maybe one of them would be a mathematician.

That first night, when I was still recovering, he ordered me to lie down, wrapped me up in a soft quilt, and wound up his gramophone. The soulful notes of Louis Armstrong's trumpet, of all things, drifted into the Delhi evening. Vijay, it turns out, has become a fan of jazz. His musical taste is mostly attributable to Mr. Hanks, the transcriber of our jailhouse conversations. He had an apartment over Vijay's cell and his nightly jamming had at first irritated and then charmed Vijay. Unbeknownst to me, Hanks had presented several gramophone recordings to Vijay before he left Morisette, which likely made Vijay the only jazz aficionado in India.

After a few days life settled into a routine. In the mornings we would venture to some fort or historical monument in Delhi, and in the late afternoons we would sit under the Banyan tree and talk. We talked about the Indian freedom movement, the latest goings-on in Morisette, and about jazz. In the evenings, he would do mathematics, and I would read. It wasn't until my eleventh day in Delhi that our conversations turned to philosophy, and then only on my steering.

"Don't you worry about certainty any more Vijay?" I had asked.

He was sitting on the ground resting his back on the bark of a mango tree and at my question he stood up and began pacing, just as he had done in the Morisette jail cell.

"I don't think it's a question we can make progress on, Judge. Certainty, at least the kind of absolute certainty you and I sought, seems utterly beyond human grasp. But I think I've learned how to deal with it."

"How do you deal with it?"

"I have faith that what my mind assures me to be true is, in fact, true. I don't demand an axiomatic proof anymore. I've let that go. I'm okay taking things on faith."

I still didn't know what exactly had turned it around for him. Who was his Darrel Huston? Who had convinced his soul that it was possible to let go?

"It was set up by Ramanujan's letter and completed with the conversation we had on your return from Dogstown," said Vijay. "I could finally step out of the trap of trying to formally justify mathematics via axioms."

Vijay once again told me that in the depths of his Eddington-induced despair, he had received a letter from Ramanujan, which challenged him to a problem that had befuddled many other mathematicians. To get his mind off his philosophical crisis, he pursued a solution, and after several hours of intense work found a proof. He was delighted, just as he always had been after solving a difficult problem. But then he asked himself what his jubilance could possibly mean in the context of formal axiomatic theory. After all, wouldn't the axioms underlying the steps of his proof of Ramanujan's problem be as suspect as Euclid's fifth postulate?

"That was when I realized, Judge," he said, "that my proof was enough. In no sense was I absolutely certain of the proof to the point of justifying, or even knowing, the underlying axioms, but my mind perceived my solution to be correct, and that was all I really needed. I see the chair you're sitting on and do not doubt that the chair is, in fact, there; it's the same with proof. Our minds have the capability to see mathematical truth, and when we see it, we should trust it. I acknowledge that this trust is purely a matter of faith. But after our conversations in Morisette, I am less inclined to fret over taking something on faith."

I asked him what problem of Ramanujan had so enamored him that night. He nodded, as if he had been expecting my question, and pulled out a worn folder from his bookcase. He pointed to an odd-looking equation in what must have been Ramanujan's handwriting:

$$x = \sqrt{1 + 2\sqrt{1 + 3\sqrt{1 + 4\sqrt{1 + \ldots}}}}$$

"Ramanujan challenged me to determine the value of x," said Vijay.

Ramanujan's equation was unlike anything I had seen before. There were square roots within square roots that went on forever. The complexity was maddening, yet the pattern was beguiling. You started with 1 and added 2 times something, where the something was like the original expression, only instead of 2, you began with 3, and on you went forever.

"I was able to show that $x = 3$," said Vijay. "The proof is not too hard, and if you want we can discuss it later, but suffice it to say, that once I constructed it, I knew the equation to be true. The grace of the result was enough for me, and thoughts of what it actually meant in an epistemological sense seemed more removed."

As gently as I could, I asked him if he now believed in God.

He smiled and shook his head. "No, Judge, that is not in my nature." He then paused, looked me in the eye, and said, "But I can understand why someone might."

A few day later, as my ship departed the Bombay harbor, I watched Vijay wave goodbye in what more likely than not would be the last time we would see each other. Just before I boarded the ship he had told me without preamble or ceremony that I was his only abiding friend. I had wanted to reciprocate *in kind, but my native reservation made me hesitate for a second and in that moment my opportunity was gone. Despite my regrettable silence I do believe, or at least hope, that he sensed the admiration and affection I have for him. We shook hands, and in a gesture that was awkward for both of us, we briefly hugged.*

Looking at him from the vantage of the ship's deck, I felt the true scope and extent of our human freedom. Here was a man who, inspired by a snake with a girl's head, had decided to understand mathematics. He had used it to develop a philosophical basis for his life, only to see that basis shattered. Yet he had chosen to regain his balance. In doing so, he had renewed his zeal for his beloved subject, and it had set him free.

A similar freedom had pervaded my life as well: I had chosen to allow the arguments of an atheist mathematician to reshape the contours of my faith. Stirred by understanding, I had chosen to free him, and the political fallout from this unpopular move had denied me my life's ambition: a seat on the Supreme Court. To this day I find that people are suspicious that my freeing the "atheist Hindu" (as several still refer to Vijay) pointed to some shifty unreliability in my own faith. Yet I remain at peace with my decision. It was my choice. Some may argue that our individual choices are already made for us and the idea of "free will" is an illusion. To them I merely reply that even if choice is an illusion, our perception of being able to choose is not. So our freedom is real—at least as real as anything else we live by.

This freedom has me in awe. It is unbounded, and every single one of us possesses it. We are free to believe or not to believe; we may create mathematics or build homes or write poetry or do nothing at all; we can marry and raise a family or stay in bachelorhood; we can quest for new adventure or find comfort in the familiar; we can seek meaning or we can doubt that it is possible to find meaning. Every path is there to be taken or ignored, and none is ordained. We are given no certainties, yet we are given the capacity to feel certainty. There is no absolute meaning to latch onto, yet transcendence is within our grasp. We are free to chart our course, free to pursue our passions, and free to create the axioms of our lives. And it is in this glorious freedom that I find grace. This freedom, then, is my proof of His existence.

• • •

Two days later, in the last class of the semester, Nico said he was going to step out of actual mathematics and discuss the philosophy of mathematics. "Some of you have already been thinking about what mathematical truth means," he said looking at Adin, "and this is one of the principal topics covered in the philosophy of mathematics."

Now, many years later, I distinctly remember Adin in front, leaning forward in anticipation.

I don't recall much of the lecture. I kept replaying John Taylor's words in my head. I finally knew what had happened to Bauji, but what did it mean? And were there any lessons for me?

I remember Nico saying that there were several schools of thought that sought to interpret mathematical knowledge. "The Platonists believe that mathematical objects exist outside the human mind, that they are discovered, not constructed, by humans, and that any intelligent aliens would come to the same conclusions."

He paused and looked at me. "Vijay Sahni, the Indian mathematician, was a Platonist when he came to Morisette." The name Vijay Sahni wouldn't have meant anything to most in the class. I was touched that Nico had used Bauji as an example.

Nico continued, "Platonism feels true to many mathematicians but it can be hard to defend philosophically. After all, one might ask where these mathematical objects exist if not in the human mind." Judge Taylor had asked exactly this of Bauji.

Nico went on to say that unlike the Platonists, the formalists believe that mathematical statements may be thought of as statements about the consequences of certain string manipulation rules. In this way of thinking, one starts with some axiom statements (which are essentially a string of symbols without any intrinsic meaning) and some rules for manipulating those statements, and ends up with theorems that are nothing more than strings in that axiom system. For example, the Pythagorean theorem is a string in the Euclidean system, and the uncountability of the real numbers is a string in the Zermelo Fraenkel axiom system. In formalism no axiom system is much better or worse than any other.

"Formalism is philosophically clean, but very few working mathematicians are real formalists. Doing mathematics does not feel like we're constructing strings in an axiom system; it feels like we're discovering new truths." I remember Nico saying this last sentence with passion. He was no formalist.

He went on to illustrate the differences in the two philosophies using the Continuum Hypothesis: To a Platonist there is a real answer out there—either there is an infinite set with cardinality between the naturals and the reals, or there isn't. The fact that the current set of axioms does not allow us to answer the hypothesis is neither here nor there; it just means that we need better axioms. To a formalist the hypothesis is a meaningless question within the current axiomatic structure. It is not a well-formed string.

There were other schools I vaguely recall from the lecture: Logicism argues that logic is the proper foundation for mathematics; constructivism insists that only mathematical entities that can be explicitly constructed should be considered valid. My favorite, however, was something call quasi-empiricism (Claire had rolled her eyes when she heard the name), whose proponents argued that truths flowed not from axioms to theorems, but rather from theorems to axioms, and the theorems themselves are discovered by trial and error and by experimentation. I liked quasi-empiricism because it didn't draw a distinction between mathematical learning and other forms of learning. Mathematics, like anything else, is seen as a human pursuit. Bauji left Morisette as a quasi-empiricist.

After class Nico invited me to his office. "Let's talk," he said. "Sit, sit." He headed to the table in the corner to brew a fresh cup of the java he seemed to live on. When it was finished brewing, he carried his cup to the desk and eased himself into his chair. I can still recall the moment in precise detail. A ray of sunlight slanted across the room, illuminating the play of dust particles in the air and lighting up the rich color of his coffee.

Two entirely disparate futures lay before me, and I felt strangely disembodied from the events unfolding in the room. Now that I look back, I can recall feeling this way only a few times in my life. The first was when I lit Bauji's pyre; the others still lay in the future.

I thought that Nico would want to discuss my future (graduate school vs. Goldman), and I wanted to do so myself. But instead I ended up talking about Bauji. Nico didn't seem to mind—he had no pressing agenda. I quickly related all that I had found in the most recent transcripts and then laid out what was bugging me. "Nico, I know what my grandfather would have wanted me to do, but I am no longer sure how I should decide this issue, or decide anything for that matter."

I had the feeling that Nico could sense the struggle going on within me and he now sounded less sure than he had been in the message he had left on the machine, "Look Ravi, I cannot answer for your grandfather or for Judge Taylor. But from my own experience I can tell you what I know and think."

"The common ground that your grandfather and the judge seem to have arrived at seems separated only by how they see themselves. Or so it seems to me. I personally have very little difficulty with either point of view. Yes, I am a believer but that may only be a question of whether my own nature prefers to affirm or deny."

You must realize how much the judge sacrificed in reaching the point of view he had. His version of faith, and it is a version that I share, leaves no room for miracles or for men walking on water or for virgins giving birth. He believed in an order that pervades the universe, and after all, isn't that a vision that in reality your grandfather shared? And not just your grandfather; I think every scientist shares that vision, otherwise why would we even attempt to make sense of the world or look for patterns and laws that guide us? It is true that absolute certainty may lie outside our reach, but we live for that magic moment of discovery when we are attuned to this sense of order and connectedness. And the existence of this order and connectedness is a leap of faith.

"Ravi, in my opinion every person should choose a life path that allows himself to nurture this sense of order and connectedness. That's what your grandfather did, and that's what—in his way—Judge Taylor did as well, and that's what I think you should do."

• • •

Just as I came back to my room, the phone rang. Fearing a call from Stella Channing of Goldman, I had deliberately let the phone go unanswered that week. She had left three messages already, and her final one said that my offer might be revoked if they didn't hear from me by Friday. Friday was only 36 hours away, and I still had no clear idea about what I would tell her, which is why I had been avoiding the phone. This time, though, distracted by my conversation with Nico, I picked it up on reflex, and sure enough it was stella.

"Ravi, have you been away somewhere?"

"Yes," I lied.

"I knew it," she said happily. "That explains why I hadn't heard from you. Listen, I know you were very excited about our offer, so we were surprised you hadn't called with your acceptance. I'd even left you a couple of messages. Anyway, to send you the next set of paperwork I just need you to formally confirm that you are in fact accepting our offer."

I didn't say anything. I still wasn't sure what to say.

"Ravi, you are accepting, aren't you?" she asked.

I had imagined this moment and had hoped that certainty would come when I needed it to come. But no, I still couldn't get anything out. What was my nature? What did I really want? What was my free choice?

"Hello? *Hello*?" said Ms. Channing.

Epilogue Epilogue Epilogue Epilogue

TIME PASSES. Decisions get made. Our hand is forced by circumstance, or impulse. We decide, and then we move on; new choices arise, new situations challenge us, and we rush onward to confront them. Our brains have evolved to interpret our decisions and to assign meaning to them.

So what meaning should I attach to choosing a life of mathematics? Should I tell myself that it represents the reawakening of my mathematical interest, first sparked by Bauji? Or was it driven by a quest to understand the nature of true knowledge? Or was it simply that mathematics meant the possibility of Claire?

Truth is, I don't know. It's been a good life, but unlike Bauji, I can envision inhabiting other lives: Peter's intensely paced banking, or Adin's ruminative travels as a jazz critic (he still plays, but mostly at home). I can imagine, without much of a stretch, doing their jobs and being reasonably happy. But despite my unwillingness to attach an exalted significance to my career path, my work has kept me engaged. After Nico, I went on to do graduate work in analytic number theory and then began teaching. My research has not been groundbreaking, but it has interested a few people, including, I'm happy to say, myself. Slowly, though, my focus has shifted to finding ways to teach mathematics so that it is possible for students to see the subject's splendor and feel the excitement of chasing problems.

This semester I've been asked to teach a course in introductory logic. The centerpiece of the class is Gödel's theorem, which informally states that all consistent axiomatic formulations of number theory include undecidable propositions. Gödel proved that if you start with any axiom system (that is not completely trivial) you will always find statements that are impossible to prove or disprove. Note that this has nothing to do with our capabilities to find a proof; instead, it bars the very existence of proofs for certain statements.

Euclid's fifth postulate and the Continuum Hypothesis are concrete examples of important, interesting questions that were shown to be undecidable in the context of their respective axiom schemas. Gödel provides a measure of finality to the conclusions we came to from the specific instances of geometry and set theory. No matter what the axiom system, truth will outrun proof.

I look at the Texas Instruments calculator on my desk and wonder how Bauji, or Nico, would have approached the teaching of Gödel's marvelous but difficult result. I feel sure that they would have found a way to reduce it to bite-sized chunks, each of which would have been meaningful in its own right.

It's a tall order and there is a lot of work to do. But today is Saturday and Claire wants to go out and get curtains for our kitchen windows. Gödel will have to wait until tomorrow.

Notes & (a short) Bibliography

The following notes are categorized by chapter and by the mathematician/philosopher referred to in the text of the novel. In most cases the notes provide historical context for the (largely fictitious) journal entries, but in two instances (Zeno's paradox and the power set theorem) we've used the notes to detail a mathematical/philosophical point not covered in the novel.

This section should not be used as an independent historical reference.

Author's Note

Blasphemy Law

The New Jersey Blasphemy law mentioned in the Author's Notes is taken from *The American Atheist*, October 1986. Details of the trial may be found in Leonard Levy's *Blasphemy* (p. 508).

Chapter 1

Zeno's Paradox

As noted in the text, it is known that Zeno of Elea lived in the fifth century BC. Almost none of his original work has survived. Our knowledge of his paradoxes is derived from the (often sketchy) writings of other philosophers, who tried to—frequently erroneously—resolve his arguments. As a result, our understanding of what exactly Zeno intended to communicate is imprecise and is based on a degree of guesswork.

It is entirely possible, indeed likely, that Zeno would have considered the proof of an infinite sum of terms converging to a finite quantity to be an incorrect—or at least incomplete—resolution of his paradox. The resolution in the text assumes that uniform motion is possible, which begs the question that Zeno seems to be raising. However, our purpose in the novel was only to find an interesting way to introduce the series $2 + 1 + 1/2 + \cdots$, not to have an in-depth discussion of Zeno's paradox itself. In his fascinating book *Mathematics: The Loss of Certainty*, Morris Kline takes a similar approach (see p. 349).

The exchange between Nico and Adin below captures a more complete resolution of Zeno's paradox. It is in the spirit of R. M Sainsbury's *Paradoxes* (see discussion on pp. 16–17). We had originally included it in the text but found it to detract from the flow.

Nico seemed to be getting ready for a break, but Adin had a problem. "I have a question," he said. "What if Zeno had a different objection altogether? He seems far too smart to have been confused by infinite series. Maybe he had another reason for thinking that it is impossible to do an infinite number of M-runs."

Nico laughed. "We do not really know why Zeno thought that an infinite number of M-runs is impossible. What do you think he might have been thinking?"

Adin pushed his glasses up on his nose. "Maybe Zeno believed that the target T cannot be reached by passing through all the mid-points represented by M_1, M_2, etc. After all, T is further to the right of any of the M points. So Zeno may have been saying that going through each M cannot take one all the way to T."

Nico nodded. "Adin, many philosophers tend to agree that your objection may indeed be the one that was bothering Zeno. But there is a response to this interpretation as well. Part of the answer lies in trying to set up a correspondence between the M points and actual lengths in space. A length in space has a starting point and an end point. What is the length corresponding to the points S, M_1, M_2, M_3,. . .? Does it include T?"

There was dead silence in the room. It seemed to me that the only possible physical length corresponding to the series of unending M points had to include T. Anything short of T would mean that there are M points that have not been included in the physical length.

Adin concurred. "It does include T," he said. "T is not in the M series, yet it must belong in the physical length corresponding to the M series."

Chapter 2

Oresme

The proof that the harmonic series diverges is indeed due to Nicole Oresme, (1323–1382), who was the Bishop of Lisieux, France. This was by no means Oresme's only scientific contribution. He was a man of varied interests and did important work in subjects such as physics and economics. A brief biographical sketch of Oresme can be found in *A Source Book of Mathematics 1200–1800*, edited by D. J. Struik (pp. 134, 135). A discussion of his proof on the divergence of the harmonic series is on page 320.

The language used in the journal entry comes the discovery of modern analysis that dates to the 19th century. Oresme would not have used terms such as "convergence," but the terminology is used here to ensure consistency with the mathematics in the rest of the text.

Further, the student Seabastien is entirely fictitious.

Chapter 3

Bhaskara

One of the last great names in the Indian mathematical tradition was Bhaskara (1114–1185), or Bhaskaracharya. He led the astronomical observatory of Ujjain, sometimes referred to as the Indian Greenwich. He authored the book *Lilavati* (the beautiful), a poetical treatise on arithmetical problems.

Lilavati came to be known in the Western world after it was translated into Persian, 400 years after it was written. The translator also noted a legend that had grown around the work. Bhaskara had cast his daughter Lilavati's horoscope to determine an auspicious time for her marriage. He realized there was only one auspicious moment when she could be married. To ensure the moment would not pass unnoticed, he constructed a simple water clock consisting of a cup floating on a tub of water. A hole at the bottom of the tub was to drain the tub in such a fashion that the cup would touch the base at the auspicious moment. Curious to see how this clock worked, Lilavati leant over the tub to examine the mechanism when, unknown to her, a pearl from her necklace fell into the water, blocking the hole. The hour for marriage passed and the legend states that Bhaskara wrote the manuscript to console his unwed daughter.

There is, however, little evidence to back the details in the legend. Kim Plofker, then of the Department of History of Mathematics at Brown

University explored the issue in her lecture "The Mathematics Textbook and the Disappointed Daughter: History of a Mathematical Urban Legend" (with E. Allyn Smith) at the Joint Mathematics Meetings of the American Mathematical Society (AMS) and the Mathematical Association of America (MAA), Baltimore, MD, January 17, 2003. A short summary of the talk is available at http://www.ams.org/ams/plofker-jmm2003.html.

Moreover, only the statement of the Pythagorean theorem given in this text and then credited to *Lilavati* appears in Bhaskara's writing. It is commonly stated that he drew the figure in the dissection proof with only the word "Behold!" by way of explanation, but the origin of this tale is unknown. There are actually no diagrams in the *Lilavati*.

Pythagoras

Pythagoras (570–490 BC) is one of the most important pre-Socratic figures in Greek philosophy. Unfortunately none of his writings survive, and the letter to Pherekydes (who indeed was his teacher) is entirely fictional, but it does remain true to what is known about the Pythagorean view of the world. The legend goes that the Pythagorean philosopher Hippasus did indeed first discover that the square root of 2 was irrational while at sea and was killed by enraged compatriots, but there is no evidence to back this claim. For some details on the Pythagorean view of the world see the discussion beginning on page 104 of Morris Kline's book.

CHAPTER 4

Cantor

The entry is apocryphal, but Cantor's differences with Kronecker are certainly not. These differences may have contributed to Cantor's feeling of increasing isolation within the mathematical world. The proof given here of the fact that the rationals are countable is due to Cantor.

Cantor was married to Maria Bohm, but as far as we know she is not known to have ever addressed Georg as a "huggable bear."

Some details of the Cantor–Kronecker conflict are provided in Amir Aczel's *The Mystery of the Aleph* (pp. 131–132 and subsequently).

CHAPTER 5

Euclid (first entry)

The entry is entirely apocryphal. Tantalos is a fictitious character and it is unknown what motivated Euclid to axiomatize geometry.

Spinoza

Baruch (Benedict) Spinoza (1632–1677) was excommunicated by the elders of the synagogue at Amsterdam for "abominable heresies." He was only 24 at the time, but his views were heretical not just for his fellow Jews but even for the Christian establishment. It was only the fact that he was born into the liberal atmosphere of 17th century Amsterdam that allowed him to keep working, but some of his most important work was only published posthumously.

The entry here, while apocryphal, reflects Spinoza's desire to do for philosophy what Euclid had done for geometry. In fact, Spinoza's major text, *The Ethics*, is modeled on Euclid's *Elements*.

Beginning with definitions such as, "That thing is said to be finite in its own kind that can be limited by another of the same nature. For example, a body is called finite because we always conceive another that is greater. Thus a thought is limited by another thought. But a body is not limited by a thought nor a thought by a body," and a collection of seven axioms that include, "From a given determinate cause the effect follows necessarily; and conversely, if there is no determinate cause, it is impossible for an effect to follow," Spinoza went on to establish propositions using the Euclidean method of proof.

Among his results is Proposition XI: God, or a substance consisting of infinite attributes, each of which expresses eternal and infinite essence, necessarily exists.

The above definitions, axioms, and propositions may be found in Spinoza's *Ethics* (p. 1 and subsequently).

Hilbert

This entry is entirely apocryphal but David Hilbert (1862–1943), one of the leading mathematicians of the modern era, was the first to realize the need for consistency of an axiom system, that is, that axioms and the consequences thereof should not contradict each other. He also went on to ask when it was possible to show that an axiom system could be complete, that is, that it could derive every truth about the area of mathematics it sought to represent. This led to the development of one of the most important streams of mathematics and much of current day computer science. Hilbert set himself the task of what turned out to be the most thorough re-examination of the Eulcidean method and the work of Euclid. His book *Grundlagen der Geometrie* published in 1899 remains a classic on the foundations of geometry.

The dictum "we must know, we shall know" dates from Hilbert's address at Königsberg on September 8, 1930. It is doubtful that he formulated the phrase in 1885, the date of the journal entry. Chapter 16 of John Derbyshire's book *Prime Obsession: Bernhard Riemann and the Greatest Unsolved Problem in Mathematics* begins with an excellent description of Hilbert's speech.

Power Set Theorem

Nico's proof assumes the axiom of choice, since it assumes the following principle which cannot be derived without that axiom: either A is greater in cardinality than P(A), or they are equinumerous, or P(A) is greater in cardinality than A. This principle is an instance of the linearity of cardinals (aka "the trichotomy principle"). Cantor's theorem can in fact be proved without the axiom of choice, but it requires use of the Schröder–Bernstein as a lemma (which states that if A is greater than or equal to B and B is greater than or equal to A, then A and B are in fact equal) whose proof, although not difficult, is too involved for this book.

Cantor (letters to Mittag-Leffler)

Gosta Mittag-Leffler (1846–1927) was a leading Swedish mathematician and a friend who stood by Cantor. Much of the work done by Cantor was published in Mittag-Leffler's journal *Acta Mathematica.* This provided an important platform for Cantor at a time when most other mathematicians were not willing to take his work on infinity seriously, let alone publish it.

The dates and the thrust of Cantor's letters to Mittag-Leffler are historically accurate. Cantor did indeed write alternately claiming to have proved the truth and then the falsehood of the Continuum Hypothesis. However, the actual text in the letters is fictitious.

Aczel's *The Mystery of the Aleph* has a description of Cantor's correspondence with Mittag-Leffler beginning on page 154.

Chapter 6

Euclid (second entry)

The entry is entirely fictitious. It is unknown whether Euclid was in any way insecure about the historical significance of the *Elements.*

There is circumstantial evidence to suggest that Euclid was hesitant in using the fifth postulate. It is true that he avoided using the postulate in the first 28 propositions of *Elements.* Kline has an excellent discussion on the issues arising out of the attempt to justify Euclid's fifth postulate (pp. 78–88). These pages also succinctly describe the developments that led to non-Euclidean geometry, and the contributions of many of the mathematicians that figure in the text are taken up here.

It also seems likely that Euclid tried to prove the fifth postulate from the other four. However, there is no historical record showing this. The theorems mentioned in the entry, while in modern notation, do capture the ideas contained in *Elements*.

Saccheri

Girolamo Saccheri (1667–1733) was an Italian and an ordained Catholic priest who was persuaded to take up the study of mathematics by Tomaso Ceva after he entered the Jesuit order at Genoa. While this entry is apocryphal, it lays out the thinking that Saccheri brought to bear on the fifth postulate. The structure and the content of the arguments presented in the entry are Saccheri's, albeit in modern notation.

As presented, Saccheri began by assuming a negation of the fifth postulate and expected that a contradiction would follow, thus showing the necessity of the Euclidean statement. But try as he might, he was not able to establish any contradiction. He shied away from making the claim that the geometry he had been led to work on was consistent, but effectively his work laid the ground for the entire edifice of non-Euclidean geometry.

Bishop Barzini, Giovanni the cook, and Carmella are entirely fictitious. As mentioned above, Ceva was indeed Saccheri's teacher. However it is unlikely if either he or Saccheri likened *The Elements* to Bach's music.

The Bolyais

Farkas (Wolfgang) Bolyai (1775–1856) and his son Janos (Johann) Bolyai (1802–1860) are among the most important figures in the drama that led to the development of non-Euclidean geometry. Farkas, a Hungarian mathematician, was a close friend of Gauss (see note on Gauss) and he personally supervised the mathematical education of his son, who showed an aptitude for the subject very early on. In fact Farkas requested Gauss to tutor his son, but Gauss declined. A few years before the birth of his son Farkas started thinking about the foundations of geometry and the independence of the fifth postulate, an interest that he shared with Gauss. When his son attempted to take up the same problem he tried very hard to dissuade him, but Janos was not convinced. He was finally able to show his father that he had managed to establish a new geometry. The work was published as an appendix to a book written by Farkas. Gauss praised Janos' work but also revealed that he had anticipated much of it. Janos saw this as a considerable setback and while he did do some important mathematical work subsequently, he was never to publish another piece of work.

Whenever possible we have included translations of the actual correspondence between the Bolyais.

In particular, the following passage from Farkas to Janos is a direct quote:

> You ought not to try the road of the parallels; I know the road to its end. I have passed through this bottomless night; every light and every joy of my life has been extinguished by it. I implore you for God's sake, leave the lesson of the parallels in peace. . . . I had purposed to sacrifice myself to the truth; I would have been prepared to be a martyr if only I could have delivered to the human race a geometry cleansed of this blot. I have performed dreadful, enormous labors; I have accomplished far more than was accomplished up until now; but never have I found complete satisfaction . . . When I discovered that the bottom of this night cannot be reached from the earth, I turned back without solace, pitying myself and the entire human race.

Further, the following passage from Janos to Farkas is also a direct quote:

> I have now resolved to publish my work on parallels. . . . I have not yet completed the work, but the road that I have followed has made it almost certain that the goal will be attained, if that is at all possible. I have made such wonderful discoveries that I have been almost overwhelmed by them, and it would be the cause of continual regret were they lost before coming to fruition. When you see them, you too will recognize them. In the meantime I can say only this: **I have created a new world from nothing**. All that I have sent you 'til now is but a house of cards compared to a tower.

Lobachevsky

Nikolay Ivanovich Lobachevsky (1792–1856) was a Russian mathematician who discovered non-Euclidean geometry independently of Bolyai and Gauss. In fact, Lobachevsky first published his work in 1829, a good two years before Bolyai's appendix was published, but due to the relative inaccessibility of papers published in Russia, Bolyai came to know of this work only in 1848. Lobachevsky was an outstanding mathematician who became a professor at the age of 23 and was appointed rector of the University of Kazan at the age of 34.

Throughout his university career he carried a heavy administrative load but yet found time to do mathematical research at the highest level. The diary entry here is apocryphal but remains true to his methods, which are very similar to the efforts of Saccheri and Bolyai. He begins, as they did, by assuming a different version of the parallel postulate and then uses it to produce a consistent geometry.

He did indeed die blind and sick and was in fact recognized for his wool processing machine. The theorems in the journal entry capture the spirit of Lobachevsky's work, although they are in modern notation.

The logical equivalence of Euclidean and non-Euclidean geometries

Vijay mentions that his argument glosses over details of defining distance in the model. The details are covered in Courant and Robbins' *What Is Mathematics?* The model described by Vijay is due to Klein. Another excellent discussion of some of the issues related to the development of non-Euclidean geometry, as well as detailed descriptions of the models for such geometries, can be found in Chapter 2 of Roger Penrose's *The Road to Reality*.

Chapter 7

Gauss

Johann Carl Friedrich Gauss (1777–1855) is ranked as the greatest of mathematicians. He contributed to almost every branch of mathematics and was instrumental in laying the foundation for many new areas of mathematical research. This note in no way seeks to describe Gauss' career; we will restrict ourselves to his contribution to the matters discussed in the text.

It does now seem true that while Gauss anticipated much of the work that was carried out by Bolyai and Lobachevsky, his treatment of Bolyai remains one of the less fortunate episodes of Gauss' career. Since he had chosen to keep his work on geometry out of the public gaze, perhaps fearing the controversy it could give rise to, it was churlish for him to grudge Bolyai the recognition he deserved as an outstanding young mathematician. For Bolyai the knowledge that his work may have covered the same paths that Gauss had traversed earlier was a blow from which he never recovered.

Gauss' experiment on the peaks of Hohenhagen-Inselberg-Brocken is factual though his motivation may have had a different context than the one provided in the text. Also factual is Gauss' choice of topic for Riemann's presentation (see p. 85 of Kline).

The rest of the entry here is apocryphal and is meant to set up the next step in the development of non-Euclidean geometry that was ushered in by Riemann.

Riemann

It was not for nothing that Gauss sprung a surprise by selecting the third of three topics submitted by Bernhard Riemann (1826–1866) for his probationary lecture at the University of Göttingen. At the time Riemann was 28 years old. All the mathematical work that Riemann did in his short life was of the highest quality. While the note here is apocryphal, it is true to the ideas that were presented by Riemann in the lecture titled, "On the Hypotheses that Lie at the Foundations of Geometry." A short description of the importance of Riemann's habilitation lecture is given on pages 126–131 of John Derbyshire's book, *Prime Obsession*.

Einstein

The lecture by Riemann, noted above, laid the groundwork for the study of the geometry of curved surfaces. As stated in this journal entry by Einstein, almost 50 years after its presentation, Riemann's work provided the necessary tools for Einstein to build upon his ideas on gravitation and space. The distance metric mentioned in the text was indeed used by Einstein in developing his General Theory.

Chapter 8

Eddington (newspaper report)

Eddington's expedition, as recorded in this newspaper report, provided one of the major experimental confirmations of Einstein's theory of general relativity. The results of the expedition made front-page news in major newspapers across the world. It was not often that the media reported on science in this fashion, but at the end of the First World War, a tired Europe was on the lookout for some good news and an expedition led by an Englishman to confirm the theory of a German was just what was needed. The headline here is modeled on a November 7, 1919 report in *The Times* that declared: "Revolution in Science, New Theory of the Universe, Newtonian Ideas Overthrown." Two days later the *New York Times* stated, "Lights all askew in the heavens / Men of science more-or-less agog / Einstein theory triumphs."

Bibliography

This bibliography is not intended to be comprehensive. It is only meant to supplement some of the details provided in the notes above. In addition to the books mentioned below, we consulted a large number of mathematical texts that have not been included in this list.

Aczel, Amir, *The Mystery of the Aleph*, Washington Square Press publication of Pocket Books, a division of Simon & Schuster, Inc., New York

Courant, Richards and Herbert Robbins, *What Is Mathematics?: An Elementary Approach to Ideas and Methods*, 2nd ed./revised by Ian Stewart, Oxford University Press, New York

Derbyshire, John, *Prime Obsession: Bernhard Riemann and the Greatest Unsolved Problem in Mathematics*, Joseph Henry Press, Washington, D.C.

Kline, Morris, *Mathematics: The Loss of Certainty*, Oxford University Press, New York, 1980

Levy, Leonard, *Blasphemy*, The University of North Carolina Press, Winston-Salem, 1995

Penrose, Roger, *The Road to Reality: A Complete Guide to the Laws of the Universe*, Jonathan Cape, London

Sainsbury, R.M., *Paradoxes*, Cambridge University Press, Cambridge, UK, second edition

Spinoza, Benedict, *Ethics*, Penguin Books, London

Struik, D.J., *A Source Book in Mathematics 1200–1800*, Princeton University Press, Princeton, New Jersey

Acknowledgments Acknowledgments

Gaurav would like to thank his wife Ritika for creating a happy home that enabled the writing of this book—this when she was caring for their young son Vir (who we hope is moved to read this book some day).

Hartosh would like to acknowledge and thank his NYU advisor Sylvain Cappell, and his number theory teacher, Harold Shapiro. Gaurav thanks his Purdue University math teachers Richard Patterson and C. D. Aliprantis (who inspired the name and the spirit of Nico) and Stanford's (late) Robert Floyd.

We both owe a significant debt to Vickie Kearn for being the best editor we could have hoped for. She believed in the book from the day she saw it and helped us focus its message and vision. Special thanks also to our readers, Dr. Alexander Paseau and Dr. Joan Richards. Their careful readings and invaluable suggestions have made this a better book. Thanks also to our copy editor—Beth Gallagher—who has an eagle-eye for catching mistakes, inaccuracies and inconsistencies.

A large number of friends read and encouraged us, and this book would not have happened without their collective help. Gaurav Bhatnagar and Punya Mishra were school friends who helped us fall in love with mathematics and were around many years later to provide invaluable suggestions as the book began to take shape. Patricia Kimball guided the book's birthing process and was an inspiration for us to get started. Mariea Datiz believed in the book even before we did. Dr. M. M Chawla, Ranjan Das, Siddhartha Deb, Chitvan Gill, Sumit Gupta, Priya Hattiangiadi, Vatsala Kaul and Dr. Chaman Nahal read drafts of the book and provided invaluable input.

Finally, both of us would like to acknowledge our intellectual debt to Morris Kline. His writings on the nature of mathematical truth aroused our curiosity and inspired us to write this book.